Springer Proceedings in Earth and Environmental Sciences

The series Springer Proceedings in Earth and Environmental Sciences publishes proceedings from scholarly meetings and workshops on all topics related to Environmental and Earth Sciences and related sciences. This series constitutes a comprehensive up-to-date source of reference on a field or subfield of relevance in Earth and Environmental Sciences. In addition to an overall evaluation of the interest, scientific quality, and timeliness of each proposal at the hands of the publisher, individual contributions are all refereed to the high quality standards of leading journals in the field. Thus, this series provides the research community with well-edited, authoritative reports on developments in the most exciting areas of environmental sciences, earth sciences and related fields.

Soufiane Haddout ·
Priya Krishnamoorthy Lakshmi ·
Antonio Mubango Hoguane
Editors

Climate Change and Ocean Renewable Energy

Springer

Editors
Soufiane Haddout
Department of Physics, Faculty of Science
Ibn Tofail University
Kenitra, Morocco

Priya Krishnamoorthy Lakshmi
Department of Civil Engineering
TKM College of Engineering
Kollam, Kerala, India

Antonio Mubango Hoguane
Centre for Marine Research and Technology
Eduardo Mondlane University
Quelimane, Mozambique

ISSN 2524-342X ISSN 2524-3438 (electronic)
Springer Proceedings in Earth and Environmental Sciences
ISBN 978-3-031-26966-0 ISBN 978-3-031-26967-7 (eBook)
https://doi.org/10.1007/978-3-031-26967-7

This Springer imprint is published by the registered company Springer Nature Switzerland AG
The registered company address is: Gewerbestrasse 11, 6330 Cham, Switzerland

One of the main objectives of this book is to present selected papers from the 1st International Conference on Climate Change and Ocean Renewable Energy (CCORE-2022), which focuses on climate change and ocean renewable energy research. This book contains the most cutting-edge research directions and achievements regarding climate change and the environment (resources, energy, etc.). All the papers have been subjected through rigorous review process to meet the requirements of international publication standard.

Preface

The *1st International Conference on Climate Change and Ocean Renewable Energy (CCORE 2022)* was held virtually on November 5, 2022, due to the outbreak of the global pandemic (COVID-19). There are many restrictions and regulations that have been imposed on countries around the world, so the virtual conference was arranged in compliance with them. To avoid physical contact spreading the COVID-19, such restrictions were put in place. There were over 30 presentations during the online conference, represented many countries including: *Morocco, India, Croatia, Bangladesh, Philippine, UK, US, Mozambique, Singapore, Indonesian, Kazakhstan, and Turkey.*

A key feature of CCORE 2022 was the opportunity for academics and engineers to exchange ideas and present research findings related to ocean renewable energy and climate change. This conference will provide an excellent opportunity to update knowledge in these fields. There were two sessions at the conference with keynote speakers, oral presentations, and an online discussion board. In the first part, keynote speakers were each allocated 30–45 min to hold their speeches. Eleven expert researchers were invited from many countries to deliver the keynote speeches: *Dr. Noemi Vergopolan* from Princeton University-USA, *Dr. Philipp Thies* from University of Exeter-UK, *Dr. Mary Ann Quirapas-Franco* from Energy Studies Institute, National University of Singapore, *Dr. Sony Junianto* from Electronic Engineering Polytechnic Institute of Surabaya (EEPIS), Indonesia, *Dr. Zhansaya Bolatova* from Almaty University of Power Engineering and Communication, Kazakhstan, *Dr. Tri Retnaningsih Soeprobowati* from Universitas Diponegoro-Indonesia, *Dr. Gökçen Uysal*, from Eskişehir Technical University-Turkey, *Dr. Samet Öztürk* from Bursa Technical University-Turkey, *Dr. Surendran Udayar Pillai* from Centre for Water Resources Development and Management-India, *Dr. Nagababu Garlapati* from Pandit Deendayal Petroleum University-India, and *Dr. Omar Farrok* from Ahsanullah University of Science and Technology- Bangladesh. Their insightful speeches had triggered heated discussions in the both sessions of the conference. Then in the second part, faculty, scientists, and research scholars were given about 5–10 min to perform their oral presentations. Every participant applauded this conference for disseminating useful and insightful knowledge in the field of climate change and ocean energy.

The proceedings is a compilation of the accepted papers and represents an interesting outcome of the conference. Topics include but are not limited to the following areas: climate change impacts on Ocean renewable energy, renewable energy technologies (e.g., coastal and offshore wind, waves, tides, ocean currents, salinity gradient power, impact of climate change on water resources and river management, ocean thermal energy conversion (OTEC), hydro power systems (HPS), etc.), global environmental change and ecosystems management, prediction modeling and decision support tools, physical oceanography, hydrology and climate change, and other related topics. All the papers have been subjected through rigorous review process to meet the requirements of international publication standard. We would like to acknowledge all of those who supported CCORE 2022. The help and contribution of each individual and institution were instrumental in the success of the conference. In particular, we would like to thank the organizing committee for its valuable inputs in shaping the conference program and reviewing the submitted papers. We sincerely hope that the CCORE 2022 turned out to be a forum for excellent discussions that enabled new ideas to come about, promoting collaborative research. We are confident that the proceedings will help create new technologies not only through scientific and engineering discoveries, but also by providing unique research references.

The Committee of CCORE 2022

Committee Members

Honor Committee

Prof. Mike Elliott is Director of International Estuarine & Coastal Specialists (IECS) Ltd. and also Professor of Estuarine and Coastal Sciences at the University of Hull, UK. He was Director of the former Institute of Estuarine & Coastal Studies (IECS) at the university from 1996 to 2017. He is Marine Biologist with a wide experience and interests, and his teaching, research, advisory, and consultancy include estuarine and marine ecology, policy, governance, and management. Mike has published widely, co-authoring/co-editing 20 books/proceedings, and >300 scientific publications. This includes co-authoring 'The Estuarine Ecosystem: ecology, threats and management' (with DS McLusky, OUP, 2004), 'Ecology of Marine Sediments: science to management' (with JS Gray, OUP, 2009), and 'Estuarine Ecohydrology: an introduction' (with E Wolanski, Elsevier, 2015). He was Editor and Contributor to the 'Coasts and Estuaries: the Future' (Wolanski, Day, Elliott and Ramachandran; Elsevier, 2019), Fish and Fisheries in Estuaries (Whitfield, Able, Blaber & Elliott; Wiley, 2022), and the Treatise on Estuarine & Coastal Science (Eds.-In-Chief - E Wolanski & DSMcLusky, Elsevier). He has advised on many environmental matters for academia, industry, government, and statutory bodies worldwide. Mike is past-President of the international Estuarine & Coastal Sciences Association (ECSA) and is now Vice-Chair of Future Earth Coasts and Co-Editor-in-Chief of the international journal Estuarine, Coastal & Shelf Science; he currently is or has had Adjunct Professor and Research positions at Murdoch University (Perth), Klaipeda University (Lithuania), the University of Palermo (Italy), Xiamen University (China), and the South African Institute for Aquatic Biodiversity. He was awarded Laureate of the Honorary Winberg Medal 2014 of the Russian Hydrobiological Academic Society. He is also Member of many national and international committees linking marine science to policy.

Prof. David Bowers is Emeritus Professor of Physical Oceanography at Bangor University in the UK. He studied Physics at Manchester at the tail end of the 1960s and came to oceanography as a mature student during an oil crisis in the mid-1970s. He obtained his Masters and PhD degrees at the University College of North Wales and worked on Postdoc at Flinders University in Australia before taking up a post as Lecturer at Bangor. He retired in 2016 as Head of physical oceanography at Bangor with a career researching tides, turbulence, particle flocculation, and marine optics. He has written several books and is currently finishing a popular science account of the oceanography of shelf seas.

Organizing Committee

K. L. Priya	TKM College of Engineering, India
S. Haddout	Ibn Tofail University, Morocco
Joan Cecilia C. Casila	University of the Philippines Los Baños, Philippines
I. Ljubenkov	Water Development, Croatia
A. M. Hoguane	Eduardo Mondlane University, Mozambique

Scientific Advisory Board

D. G. Bowers	Universite of Bangor, UK
D. Abessa	São Paulo State University (UNESP), Brazil
F. Matic	Institute of Oceanography and Fisheries - Split, Croatia
V. Baiju	TKM College of Engineering Kollam, India
K. R. Renjith	Centre for Water Resources Development and Management, India
C. C. Boaventura	Eduardo Mondlane University, Mozambique
M. A. Quirapas-Franco	Aquatera Asia Pte Ltd., Singapore
S. Junianto	Electronic Engineering Polytechnic Institute of Surabaya, Indonesian
Z. Bolatova	Almaty University of Power Engineering and Communication, Kazakhstan
Ronaldo B. Saludes	University of the Philippines Los Baños, Philippines
Y. Haddout	Ibn Zohr University, Morocco
A. Rafiki	Ibn Zohr University, Morocco
E. Essaghir	Hassan II University, Morocco
D. Elemam	Damietta University, Egypt
Decibel V. Faustino-Eslava	University of the Philippines, Los Baños, Laguna, Philippines
G. T. Gimiliani	São Paulo State University (UNESP), Brazil
Garlapati Nagababu	Pandit Deendayal Energy University, India
Tri Retnaningsih Soeprobowati	Universitas Diponegoro, Indonesia
Gökçen Uysal	Eskişehir Technical University, Turkey
A. Haddout	Ibn Tofail University, Morocco
S. Mesbaholdin	Islamic Azad University, Iran

Sponsors

UNIVERSIDADE
EDUARDO
MONDLANE

AEIC
让学术交流更简单
Make Academic Exchange Easier

Contents

Development and Utilization of Ocean Renewable Energy

Zero Carbon Emission Based Electrical Power Plant by Harvesting Oceanic Wave Energy: Minimization of Environmental Impact in Bangladesh

Selim Molla[1,3], Omar Farrok[2(✉)], and Mohammad Jahangir Alam[1]

[1] Department of Electrical and Electronic Engineering, Bangladesh University of Engineering and Technology, ECE Building, West Palashi Campus, Dhaka 1205, Bangladesh
eee.n_selim@yahoo.com, {1021064001,mjalam}@eee.buet.ac.bd

[2] Department of Electrical and Electronic Engineering, Ahsanullah University of Science and Technology, Dhaka 1208, Tejgaon, Bangladesh
omar.eee@aust.edu

[3] Shahid Abdur Rab Serniabat Textile Engineering College, Government of the People's Republic of Bangladesh, C&B Road, Barishal, Bangladesh

Abstract. Fossil fuels are mostly utilized for producing electricity in many countries including Bangladesh, which emit bulk amount of greenhouse gases. At present, about 3% electricity is produced in Bangladesh from renewable energy sources (RESs) viz. solar, wind, hydro, and bioenergy to reduce the emission level. In this paper, an oceanic wave energy based direct drive linear generator is proposed to produce electrical power. As the oceanic wave energy converter results almost zero carbon emission it can mitigate the adverse environmental impact. Mathematical model of the wave energy and linear generator are presented. The proposed generator design and its working principle are described. The generated voltage, armature current, power generation, magnetic flux linkage, and mechanical force are plotted. Load profile of the generator are also presented. High graded magnetic materials are applied to the proposed generator to reduce the conversional power loss. Simulation results show that the proposed generator produces approximately 5.44 kW of electrical power (maximum) where the peak voltage of the proposed generator is 198.5 V. Therefore, the oceanic wave farm can be constructed with several linear generators to produce sufficient electricity and mitigate negative environmental impact. As the harvesting of oceanic wave energy does not require any land area, the corresponding impact of using the land area can be avoided which is another benefit of using it. Moreover, the proposed power generation scheme can be supportive not only for Bangladesh but also for the countries that have adjacent ocean and plan to achieve net zero emission target.

Keywords: Environmental impacts · Linear generator · Oceanic wave energy · Renewable energy sources

S. Haddout et al. (Eds.): CCORE 2022, SPEES, pp. 3–22, 2023.
https://doi.org/10.1007/978-3-031-26967-7_1

1 Introduction

Environmental issues are one of the key concerns as well as significant consideration in electricity generation. Generally, electricity is produced in various countries including Bangladesh by mostly utilizing fossil fuels such as coal, natural gas, diesel etc. [1]. Fossil fuels are very harmful for our planet due to their properties of greenhouse gas emission. Nowadays, some renewable energy sources (RESs) viz. solar, wind, hydro, bioenergy etc. are being applied to extract electricity. But they also emit greenhouse gas although its quantity is lower than the emission of fossil fuel. Another newly developed renewable energy, oceanic wave energy is being harvested in different leading countries for producing electricity. It has a lower amount of environmental impact than the other RESs. In addition, the availability and energy density of oceanic wave are the highest among the RESs [2].

Already, the potential of oceanic wave energy has been estimated for harvesting purposes. But it is executed considering global perception. Sometimes, it is done based on some countries and continents. The potential of the deep water of the EU is found 320GW [3]. Generally, a wave energy converter, electrical generator, and power electronic devices or converters are required for harvesting oceanic wave energy.

Recently, direct drive linear generators (DDLGs) are being developed by scientists and researchers for harvesting oceanic wave energy. The development is conducted in various aspects such as minimization of reluctance of the magnetic core, enhance the efficiency, and reducing core loss. Reluctance of the stator of a DDLG is minimized for enhancing the output power and minimizing the core loss by modifying the stator design [4]. A linear generator is analyzed to generate electricity efficiently even at low translator speed by multi-physical coupling [5]. Besides, core loss of a linear generator is reduced by applying a high graded magnetic core [6, 7].

Generally, it is found that most of the electricity is produced by fossil fuels in the small country with high population and lower economy such as in Bangladesh [1]. Although some RESs [8] namely solar and wind energy are being utilized in Bangladesh for electricity generation and minimization of the environmental impacts. The RES based power plants require large amount of land area. But the country has no sufficient land for installing the RES based power plants. It is analyzed that the problem can be solved by utilizing oceanic wave power as Bangladesh has the world's largest continuous natural sea beach. Therefore, it has sufficient amount of energy potential which can be harvested to meet the energy demand and reduce the environmental pollution. In this paper, an oceanic wave-based power station is proposed for Bangladesh as an example of this category of the country.

In this paper, a linear generator is designed and analyzed by ANSYS/Maxwell software. The mechanical force is applied to the translator of the linear generator through a simulation environment. Mathematical model of the oceanic wave and linear generator are presented in detail. Then the generated voltage of the linear generator, magnetic flux linkage, electrical power, current flow, and mechanical force are plotted with respect to time. Additionally, magnetic field intensity and flux density are illustrated with rainbow spectrum to observe the flux distribution. From the simulation results it is found that the proposed linear generator produces enough electrical power compared to its size.

2 Challenges for Electrical Power Generation

There are two challenges found for the production of electricity worldwide. First one is the availability of fuels and the other is the environmental challenge. At present, coal, oil and natural gas are being widely applied for electricity generation in the world as shown in Fig. 1 [9]. Besides the fossil fuel several countries also utilize RESs for the same purpose.

In addition, the fuel stock is decreasing day by day. For this reason, the utilization rate of the fuels would go downward in the future as shown in Fig. 1. In spite of the fact that hydroelectricity is the oldest renewable energy system in the world, which was utilized widely. But there is no significant progress of it as it is treated to be harmful for the environment. According to the forecast [9], like hydroelectricity, the utilization rate of nuclear energy would not be increased until 2050 globally. On the other hand, utilization rate of renewable energy is rapidly increasing as observed in Fig. 1 because of its clean and environment friendly nature.

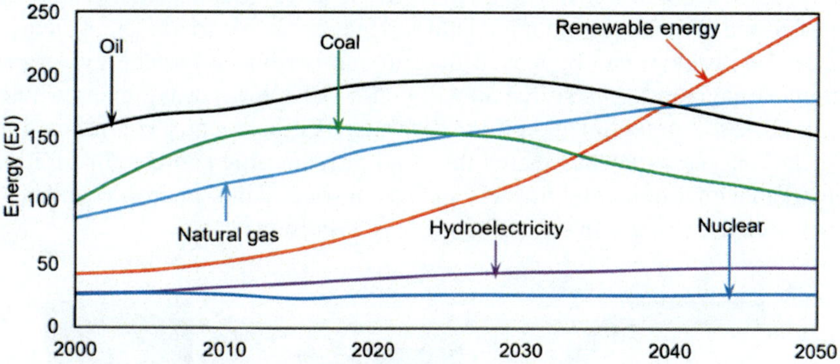

Fig. 1. Global energy consumption (2000–2050)

Nowadays, solar, wind, bioenergy, geothermal, hydropower, tidal etc. are implemented to the electricity sector in Bangladesh. Some environmental impacts are found for utilizing RESs which are mentioned in the next section.

3 Environmental Impact of the Conventional Power Plant

Most of the conventional resources are very detrimental owing to the emission of huge amount of greenhouse gases. Although the severity differs from source to source. Generally, the emission rate of natural gas is lower than both of oil and coal. However, they emit more of less amount of greenhouse gas. The electricity generation and total GHG emission in Bangladesh are plotted in Fig. 2 [10].

From here it is found that the rate of emission is increasing with the increase of electricity generation. The emission rate of natural gas is approximately half of the coal. For reducing the emission, RESs can be employed to produce electricity. But they contain comparatively lower impacts than fossil fuels. At present, solar, wind, hydro, biomass and bioenergy are utilized in the country as presented in Fig. 3 [8]. It is seen that solar energy is the highest utilized renewable source in the country.

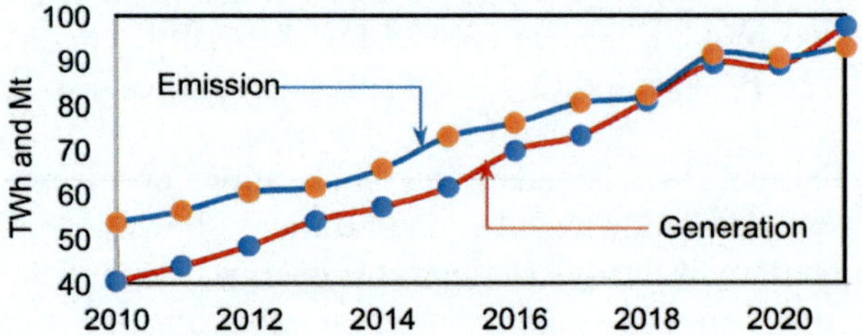

Fig. 2. Electricity generation and GHG emission in Bangladesh

The amount of GHG emission because of using coal, diesel, and natural gas is summarized in Table 1 [11, 12]. On the other hand, there is some environmental impacts for utilizing the RESs which are tabulated in Table 2. It is found from this table that some of the RESs require a huge amount of land and they are liable for different environmental impacts. The problem can be solved by utilizing oceanic wave energy for producing electricity. Bangladesh has sufficient (around 5–10 kW/m) wave energy potential as found in [20]. The potential and proposed system for the country is presented in Fig. 4. Land is not necessary for harvesting this kind of renewable sources. In addition, it has no remarkable environmental impacts such as in solar, wind, hydropower, and biomass energy which are being utilized in Bangladesh at present.

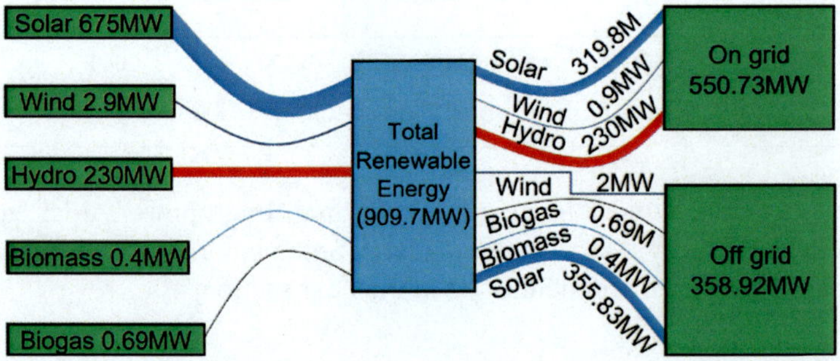

Fig. 3. Utilized RESs in Bangladesh

4 Overview of Oceanic Wave Energy Conversion

Bangladesh is a small country and there is scarcity of fossil fuels and some of the RESs as well. As it has one of the longest sea shores in the world, oceanic wave energy can be very significant candidate for the country to generate electricity, minimizing adverse environmental impact, and ensure energy security. In addition, it contains the highest

Table 1. GHGs emission for utilizing fossil fuels

Fuel	GHG emission parameter	GHG emission (gCO_2/kWh)
Diesel	Carbon dioxide	760
	Carbon monoxide	5
	Unburned hydrocarbons	0.21
	Particulate matters	0.0293
	Sulphur dioxide	1.8
	Nitrogen oxide	4.55
Coal	Carbon dioxide	900
	Sulphur dioxide	7.07
	Nitrogen oxide	4.28
Natural gas	Carbon dioxide	566
	Carbon monoxide	1.86
	Particulate matters	0.0525
	Nitrogen oxide	3.9

Table 2. Environmental impacts for using different RESs

RES	Environmental impacts
Solar	32–82 g CO_2 emits per kWh [13], high initial cost, output power variation is high [14], require large amount of land [13]
Wind	Make noise [15], hampers and kills the birds, high output power variation, low availability, emits 29.5 g CO_2 per kWh [16], visual disturbance [15]
Hydropower	Emits 31.2 gCO_2 per kWh [17], low potential, low energy density, high transmission cost, responsible to increase the temperature of local environment [18]
Biomass	Emits greenhouse gas 45.84 g CO_2 eq/MJ [14], huge amount of water is required, low energy density, requires large amount of land [19]

availability as well as energy density than the other RESs. The wave power is calculated by

$$P_{wave} = \rho g^2 A^2 T_w / 64\pi \tag{1}$$

where P_{wave} defines the power of oceanic waves, ρ is the density of ocean water, g means the gravitational acceleration constant, A denotes the wave height, and T_w is the wave energy period. At present, the US, UK, Canada, Australia, China, Portugal, Sweden, and some other countries are producing electricity from the source.

There are several types of converters for harvesting oceanic wave energy, such as point absorber, Archimedes wave swing, oscillating water column, oscillating wave

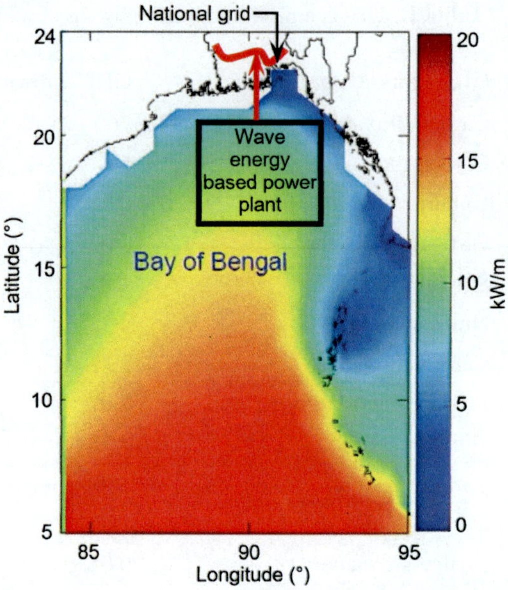

Fig. 4. Proposed power plant along with wave energy potential in Bangladesh [20]

surge, overtopping device, bulge head wave energy converter, submerged pressure differential, attenuator, and rotating mass. Point absorber is popularly applied to the conversion process. There are two types of generators applied to the converter. One is traditional rotary electrical generator and another is a direct drive linear generator. Rotational generator requires a complex gear box arrangement that drives the rotor of the generator. But the DDLG does not require complex gear box unit. Therefore, its construction is much easier than the conventional generator. For such reasons, it is popularly employed to the point absorber converter as shown in Fig. 5.

The DDLG has mainly two parts. First one is the translator as shown in the left-hand side of Fig. 5 and other one is the stator which is placed very close to the translator. Although the stator is shown in the right-hand side, but in practice, it is placed on both sides of the translator for balancing purpose. One side of the translator is joined with a floating buoy through a rod with high tensile strength and the opposite side is joined with the base through the spring. The base is basically used for mooring purpose. With the propagation of oceanic wave, the translator moves with the buoy in vertical direction. Because of the relative motion between the stator and translator, electromotive force is induced in the stator by Faraday's laws of electromagnetic induction. Then the generated ac power is converted to dc power. Because the frequency and voltage of the induced electromotive force are not the same as load or grid side power system. Further, an inverter is applied to convert the dc power into ac power to produce required voltage and frequency. Then a step-up transformer is applied to increase the voltage to the required level. Finally, it supplies the ac power to the load or grid.

In general, oceanicwave is modeled assinusoidalwaveform. Although the oceanic wave is not exactly sinusoidal, butbecause of approximation, complexity can be avoided for the simulation of linear generators. The commonlyused vertical velocity range of

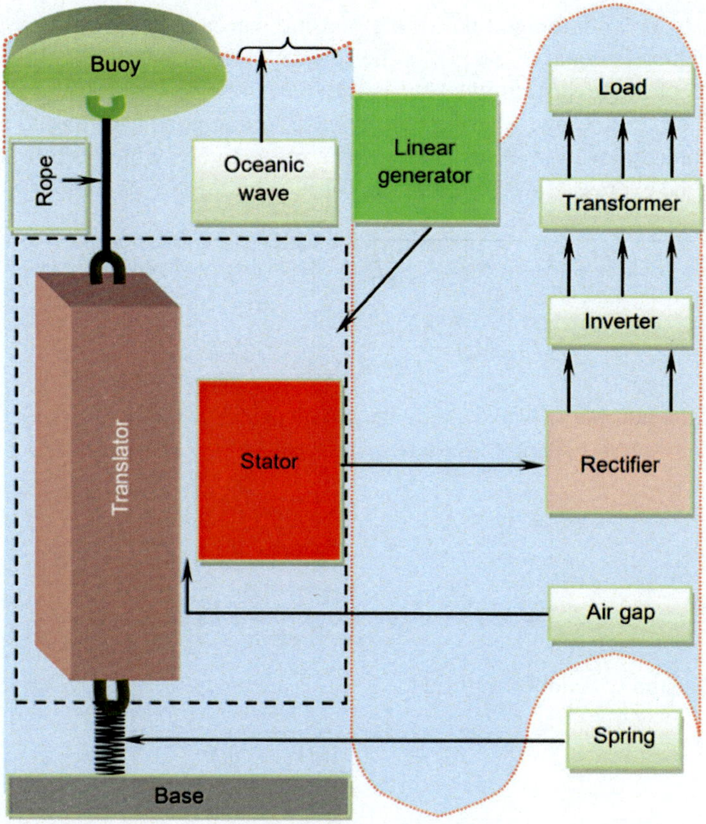

Fig. 5. Working principle of wave energy conversion with a linear generator

oceanicwaveis 0–2 m/s where the time period varies from 4–6 s. The proposed generator is suitable for installation in shallow water of the ocean. The model of the shallow water wave (free wave) with sinusoidalshapeisshown in Fig. 6.

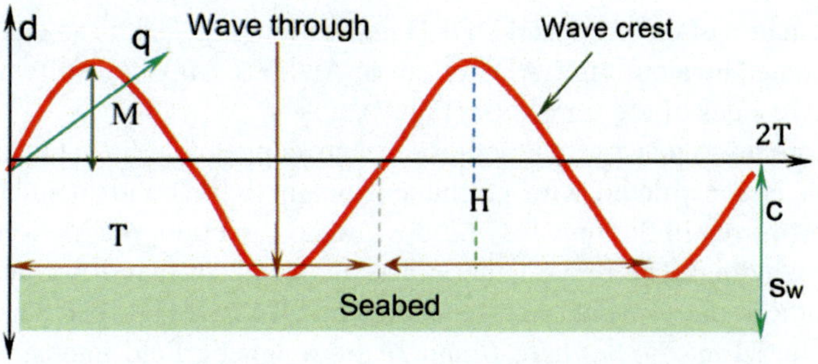

Fig. 6. Mathematical representation of the oceanic wave

In Fig. 6, M is the magnitude of the oceanic wave, H is the peak to peak value of the wave, which is equal to $2M$, cq is the surface of the ocean in the three dimensional Cartesian coordinate system where d is the elevation, S_w is the water depth from the ocean surface. It is known that for linear wave, q component is zero.C and D are the propagation and elevation of the linear wave, respectively, which is a function of time, t. They can be expressed in the following for the condition $T >> S_w$.

$$C = M_\omega \frac{\cosh\{k(d + s_\omega)\}}{\sinh(ks_\omega)} \cos(kc - \omega t) \tag{2}$$

$$D = M_\omega \frac{sinh\{\ k(d + s_\omega)\}}{sinh(ks_\omega)} sin(kc - \omega t) \tag{3}$$

where k is the number of waves, ω is the frequency, T is the time period, For the other condition T > > M, (2) and (3) can be expressed as

$$C_s = M_\omega \frac{e^{k(d+s_\omega)} + e^{-k(d+s_\omega)}}{e^{ks_\omega} - e^{-ks_\omega}} \cos(kc - \omega t) \tag{4}$$

$$D_s = M_\omega \frac{e^{k(d+s_\omega)} + e^{-k(d+s_\omega)}}{e^{ks_\omega} - e^{-ks_\omega}} \sin(kc - \omega t) \tag{5}$$

Now, (4) and (5) canbesimplified as

$$C_s \cong \frac{M_\omega}{ks_w} \sin(kc - \omega t) \tag{6}$$

$$D_s = M_\omega \left(1 + \frac{d}{s_\omega}\right) \sin(kc - \omega t) \tag{7}$$

For the condition $ks_w << 1$, $\omega = k\sqrt{gs_w}$. For $v = \frac{\omega}{k}$, the phase of the oceanic wave can be shifted as $v = \sqrt{gs_w}$.

5 Construction of the Proposed DDLG

The basic structure of the proposed DDLG is shown in Fig. 7. It is an electrical machine which is designed by a built in CAD system of ANSYS/ Maxwell software. Two stators are located two sides of the translator (Fig. 7).

The stators are made by magnetic core and copper winding. On the other hand, the translator is constructed with permanent magnet (PM) and magnetic core. When the translator moves in the upward and downward direction, magnetic flux cut occurs in the stator windings.The linear generator is analyzed for two different ferromagnetic materials, namely, the conventional steel core and DI MAX HF-10X. Their magnetizing curves are plotted in Fig. 8 where B and H are magnetic field intensity and magnetic flux density, respectively. It is found that the magnetic saturation of DI-MAX HF-10X is lower than the conventional steel core. Core loss curves of the conventional steel core and DI-MAX HF-10X cores are shown in Fig. 9 for comparison. Phasor diagram of the proposed linear generator as presented in Fig. 7 is sketched in Fig. 10.

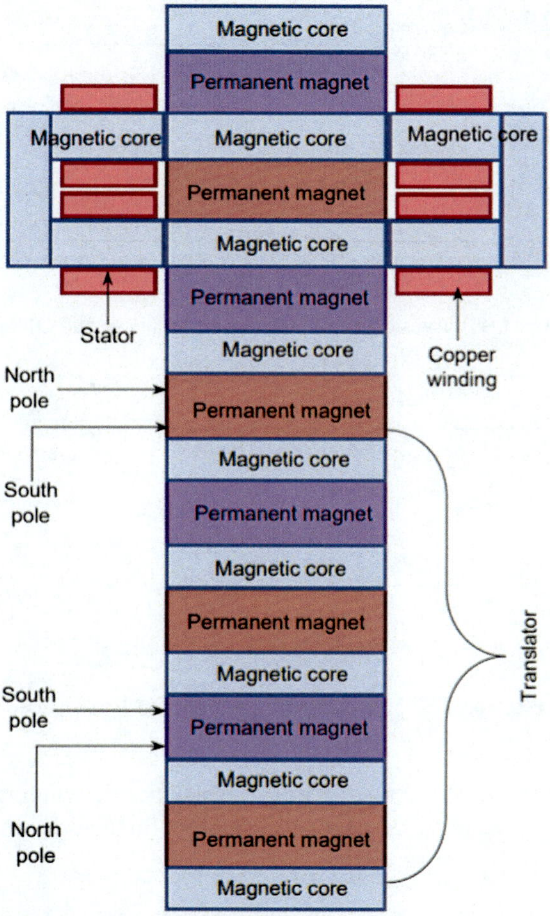

Fig. 7. Construction of the proposed linear generator

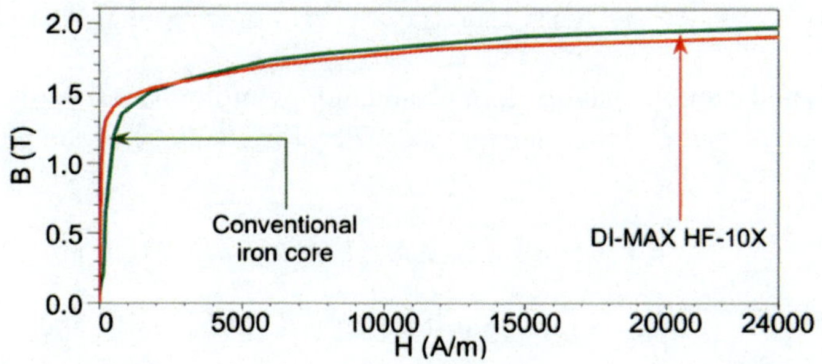

Fig. 8. Magnetizing curves of the conventional steel core and DI-MAX HF-10X

The generated voltage of the proposed linear generator is expressed in the following.

$$v = ri + \frac{d\psi}{dt} \tag{8}$$

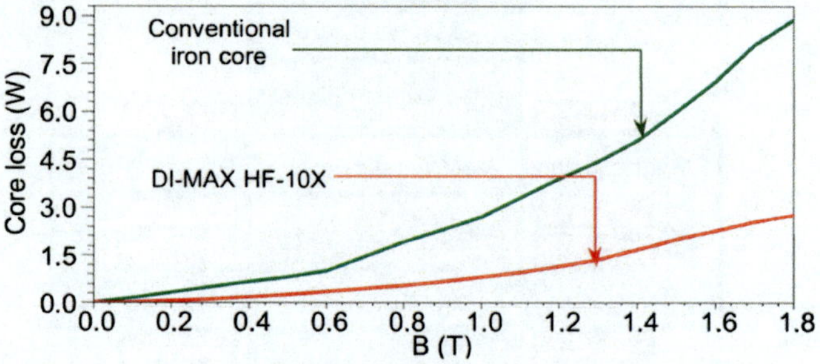

Fig. 9. Core loss curves of the conventional steel core and DI-MAX HF-10X

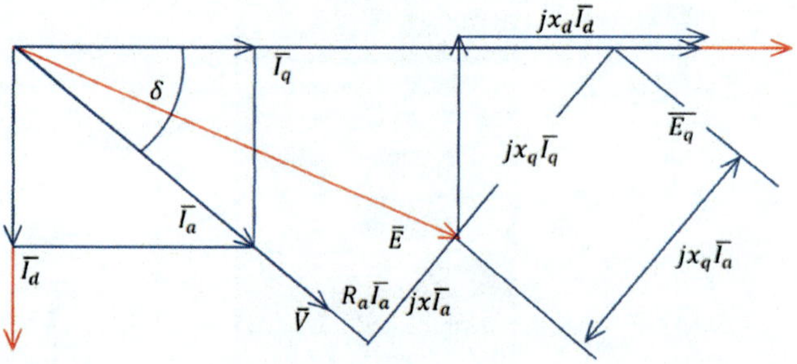

Fig. 10. Phasor diagram of the proposed linear generator

where *v*, *r*, *i*, and ψ are the applied voltage, resistance of winding, input current, and flux linkage with the winding, respectively. The flux linkages for the three phase stators can be expressed as

$$\left.\begin{aligned} \psi_a &= L_a i_a + M_{ab} i_b + M_{ac} i_c + M_{af} i_f \\ \psi_b &= M_{ba} i_a + L_b i_b + M_{bc} i_c + M_{bf} i_f \\ \psi_c &= M_{ca} i_a + M_{cb} i_b + L_c i_c + M_{cf} i_f \\ \psi_f &= M_{fa} i_a + M_{fb} i_b + M_{fc} i_c + L_f i_f \end{aligned}\right\} \quad (9)$$

where *a*, *b*, *c*, and *f* are the phases *A*, *B*, *C*, and field winding *F*, respectively. *L* is the self inductance and *M* is the mutual inductance. Therefore, voltage of the phase winding A is

$$v_a = r_a i_a + p\left[L_a i_a + M_{ab} i_b + M_{ac} i_c + M_{af} i_f\right] \quad (10)$$

where *p* is the differential operator. Substitution of L_a, M_{ab}, M_{ac}, and M_{af}in (9) yields

$$\begin{aligned} v_a &= r_a i_a + p\big[\big(L_{ao} + L_{g_2}\cos 2\theta\big) ia \\ &+(-0.5L_{go} + L_{g_2}\cos(2\theta - 240^\circ))i_c \\ &+(-0.5L_{go}L_{g_2}\cos(2\theta - 240^\circ))i_c + \big(M_{afm}\cos\theta\big)i_f\big] \end{aligned} \quad (11)$$

The two axis model can be represented by Park's transformation where the voltages v_a, v_b, and $v_c$are in terms ofv_d, v_q, and v_o, respectively which are expressed in the following

$$\begin{bmatrix} v_d \\ v_q \\ v_o \end{bmatrix} = \sqrt{\frac{2}{3}} \begin{bmatrix} \cos\theta & \cos\left(\theta - \frac{2\pi}{3}\right) & \cos\left(\theta - \frac{4\pi}{3}\right) \\ -\sin\theta & -\sin\left(\theta - \frac{2\pi}{3}\right) & -\sin\left(\theta - \frac{4\pi}{3}\right) \\ \frac{1}{\sqrt{2}} & \frac{1}{\sqrt{2}} & \frac{1}{\sqrt{2}} \end{bmatrix} \begin{bmatrix} v_a \\ v_b \\ v_c \end{bmatrix} \tag{12}$$

$$\left.\begin{aligned} v_d &= r_a i_d + p\psi_d - \omega\psi_q \\ v_q &= r_a i_q + p\psi_q - \omega\psi_d \\ v_o &= r_a i_o + p\psi_o \end{aligned}\right\} \tag{13}$$

where ψ_d, ψ_q, ψ_o are in terms of ψ_a, ψ_b, ψ_c. The terms $p\psi_d$ and $p\psi_q$ are transformed voltages. As the voltage is periodic, $\omega\psi_d$ and $\omega\psi_q$ are related to the voltage frequency. Under balanced operation, v_o can be avoided. The field circuit voltage can be expressed as follows.

$$v_f = r_f i_f + \sqrt{\frac{3}{2}} pM_{afm} i_d + L_f p i_f \tag{14}$$

$$\psi_f = L_f i_f + \sqrt{\frac{3}{2}} pM_{afm} i_d \tag{15}$$

$$\left.\begin{aligned} \psi_d &= L_d i_d + \sqrt{\tfrac{3}{2}} pM_{afm} i_f \\ \psi_q &= L_q i_q \\ \psi_0 &= L_o i_o \end{aligned}\right\} \tag{16}$$

For simplicity, let

$$\sqrt{\frac{3}{2}} pM_{afm} = M_d \tag{17}$$

Now, (16) and (17) become

$$\left.\begin{aligned} \psi_d &= L_d i_d + M_d i_f \\ \psi_q &= L_q i_q \\ \psi_f &= M_d i_d + l_f i_f \\ \psi_0 &= L_o i_o \end{aligned}\right\} \tag{18}$$

and (13) become

$$v_d = r_a i_d + L_d p i_d + M_d p i_f - \omega_r L_q i_q \tag{19}$$

$$v_q = r_a i_q + L_q p i_q + \omega_r L_d i_d + M_d \omega_r i_f \tag{20}$$

$$v_o = r_a i_a + L_o p i_o \tag{21}$$

Therefore, (14) can be expressed as

$$v_f = r_f i_f + M_a p i_d + L_f p i_f \tag{22}$$

In (19), the following parameters have been introduced with their expression.

$$\left.\begin{array}{c} L_d = l_a + \frac{3}{2}(L_{go} + L_{g2}) \\ L_q = l_a + \frac{3}{2}(L_{go} - L_{g2}) \\ L_o = l_a \end{array}\right\} \tag{23}$$

During the operation of the linear generator core loss occurs in the magnetic core, which can bedetermined as follows

$$\Delta P_{core} = \left[K_h C_h \left(\frac{f}{50}\right) B^2 + K C_e \left(\frac{f}{50}\right) B^2\right] M_{core} \tag{24}$$

where f is the voltage frequency, B is the magnetic flux density, K_h, C_h, K_e, and C_e are coefficients.

6 Simulation Results

The translator velocity is set to 1m/s speed with respect to the stators in the simulation environment. The properties of each of the material are assigned to all the components of the proposed DDLG. Force parameter is assigned for the translator. Because the force is applied to the translator for electrical power generation. Then mesh set up is applied to all the components of the DDLG as shown in Fig. 11.

Boundary setup is applied to the DDLG as a part of the finite element analysis before running the simulation. The accuracy of output parameters of the linear generator depends on all the setup values. The following simulation results are presented for a 3.5 Ω load resistance connected to the proposed linear generator. The waveform of mechanical force is plotted in Fig. 12. It is found that average 2.94 kN force is applied to the translator to produce electrical power.

The magnetic flux links with the stator windings during the operation of the linear generator, which is presented in Fig. 13. The peak value of flux linkage is 1.76 Wb. Waveform of the magnetic flux is sinusoidal. The generated voltage waveform of the proposed DDLG is plotted in Fig. 14.

First 50 ms is the transient period. In this time interval, no stable results are observed. For this reason, the results are considered from 50 ms. The peak voltage of the generator is 140.5 V where the rms voltage is 99.1 V.

When the generator is connected to the load, current flows through it. In the simulation, resistive load has been connected. The waveform of current is shown in Fig. 15 where 23.3 A current (rms) flows through the load.

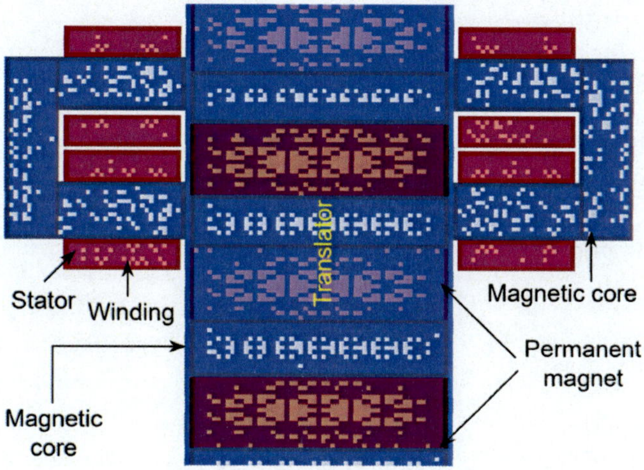

Fig. 11. Mesh setup of the proposed linear generator

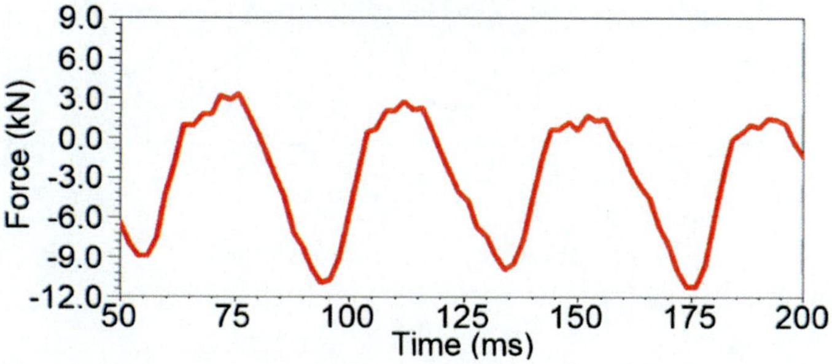

Fig. 12. Applied force to the linear generator

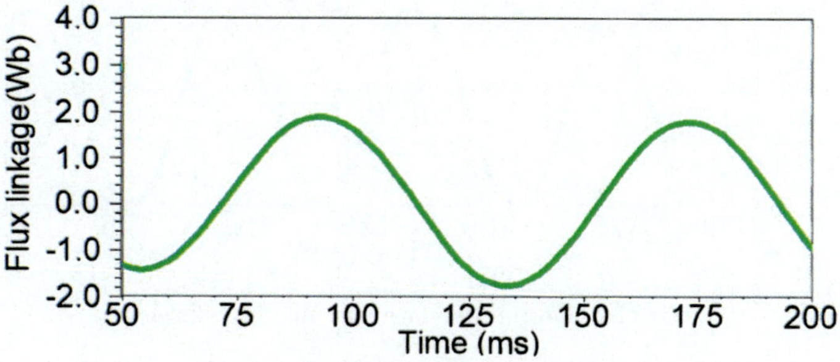

Fig. 13. Flux linkage of the linear generator

The waveform of generated power is plotted in Fig. 16. It is found that the linear generator generates 4.7 kW peak power. Load profile of the generator is shown in Fig. 17 where the maximum power is generated for 6.5 Ω load resistance. Maximum power for the ordinary steel core is 5.444 kW whereas the same for DI-MAX HF-10X core is 5.438 kW. The generated powers for different loads are tabulated in Table 3.

The generated voltage, load current, electrical power, magnetic flux linkage, and applied mechanical force to the translator are presented in Figs. 18, 19, 20, 21 and 22

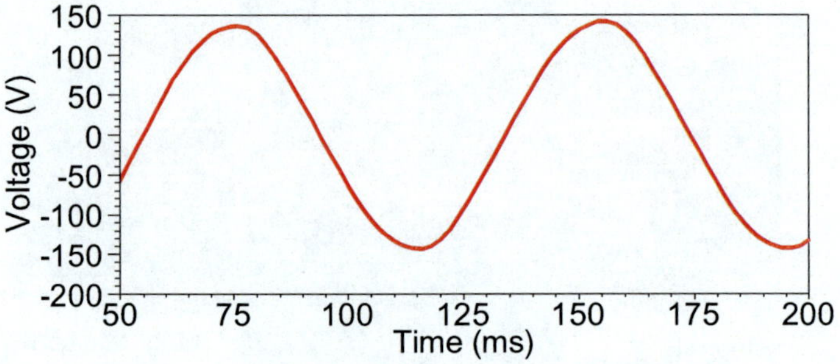

Fig. 14. Generated voltage of the linear generator

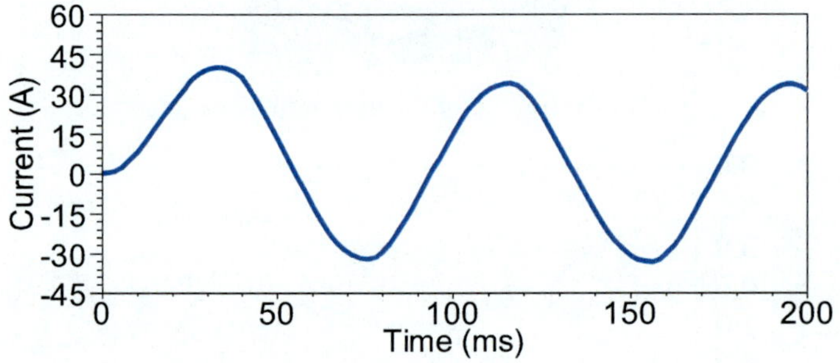

Fig. 15. Current flow through the load

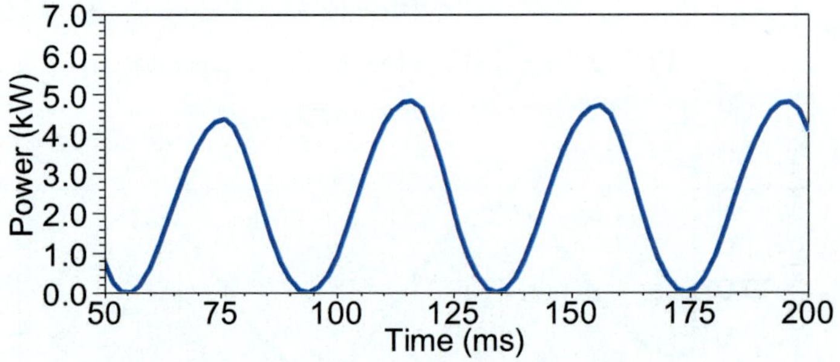

Fig. 16. Output power of the linear generator

for the maximum power (for 6.5 Ω load). As the core loss of the conventional steel core is high as found in Fig. 9, DI-MAX HF-10X magnetic core is selected for the proposed linear generator. It is found that for the maximum power generation, the peak voltage of the proposed generator is 198.5 V, current is 19.44 A (rms), average power is 2741 W, Peak flux linkage is 2.5 Wb, and average mechanical force is 3.5 kN. Magnetic field intensity and flux density of the proposed linear generator are illustrated in Figs. 23 and 24, respectively.

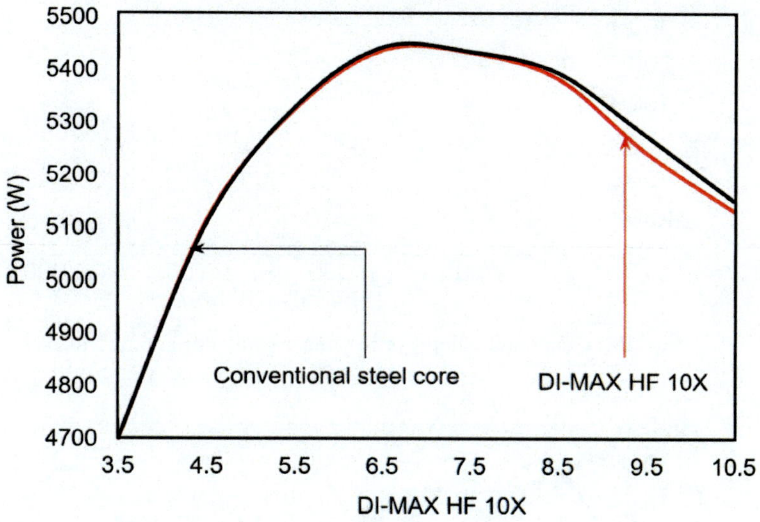

Fig. 17. Load profile of the proposed linear generator.

Table 3. Output power of the proposed DDLG for different loads with the two magnetic materials

Load resistance (Ω)	Power for using DI-MAX HF 10X (W)	Power for using conventional steel core (W)
3.5	4698	4701
4.5	5115	5111
5.5	5326	5330
6.5	5438	5444
7.5	5434	5436
8.5	5384	5396
9.5	5243	5276
10.5	5134	5154

It is observed that the magnetic field is uniformly distributed in the proposed generator. Magnetic field reaches to the highest value around the stator and translator pole faces. It is observed in Fig. 24 that magnetic flux density is the least in the copper winding. Dimension and parameters of the proposed generator are summarized in Table 4. It is found from Table 4 that the linear generator produces enough power (peak power is 5.44 kW approximately) compared to its size. It means that it is possible to generate high amount of electrical power by harvesting oceanic wave, because in an oceanic wave farm, several generators are connected.

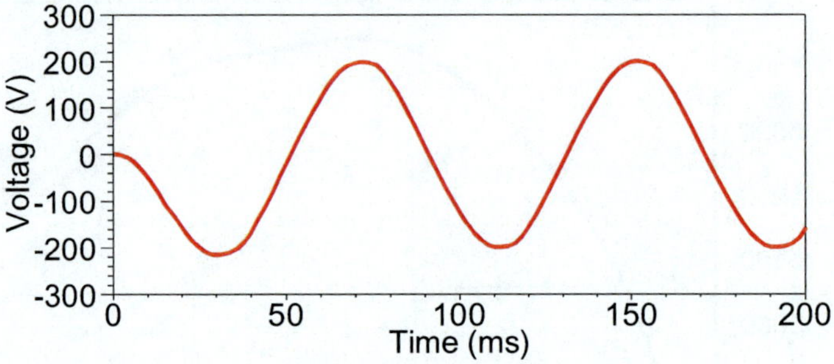

Fig. 18. Generated voltage of the linear generator at 6.5 Ω load

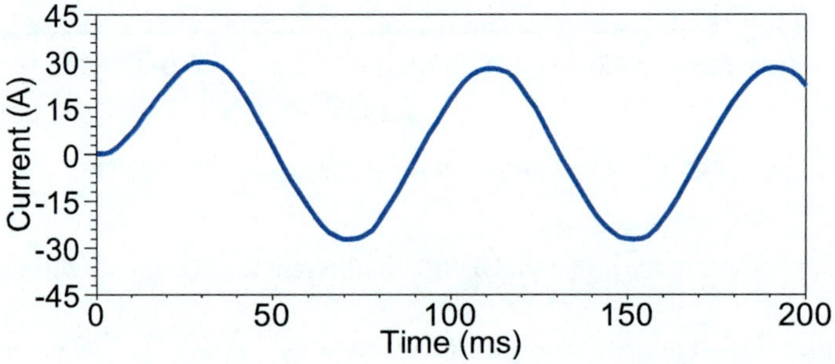

Fig. 19. Current flow through the 6.5 Ω load

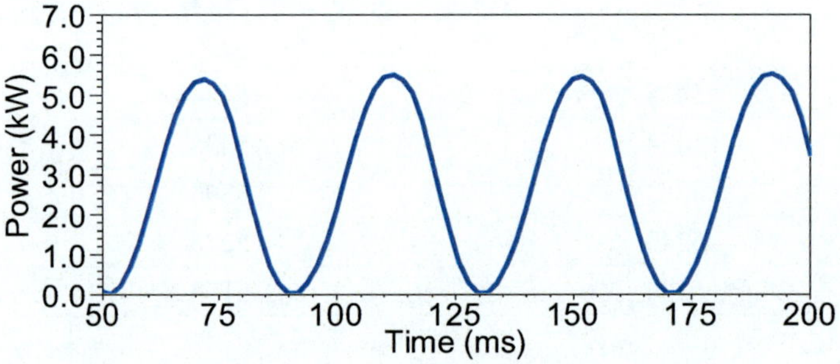

Fig. 20. Output power of the linear generator at 6.5 Ω load

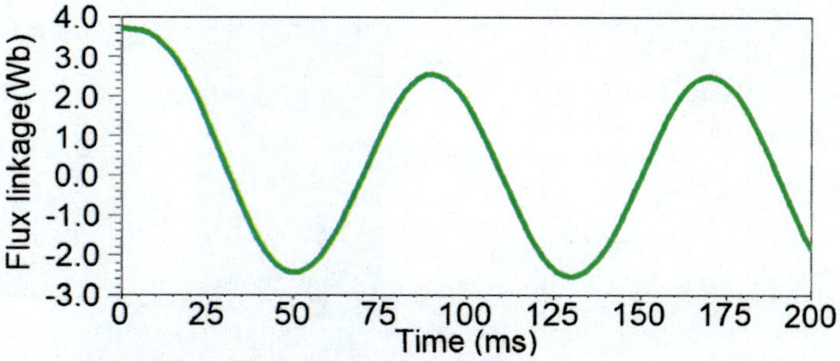

Fig. 21. Flux linkage of the linear generator at 6.5 Ω load

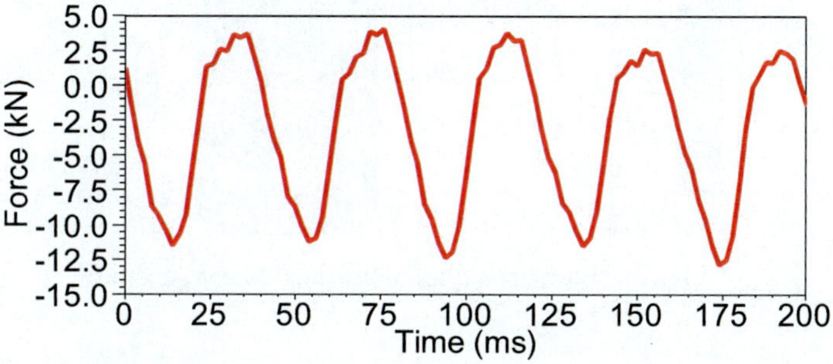

Fig. 22. Applied mechanical force to the linear generator

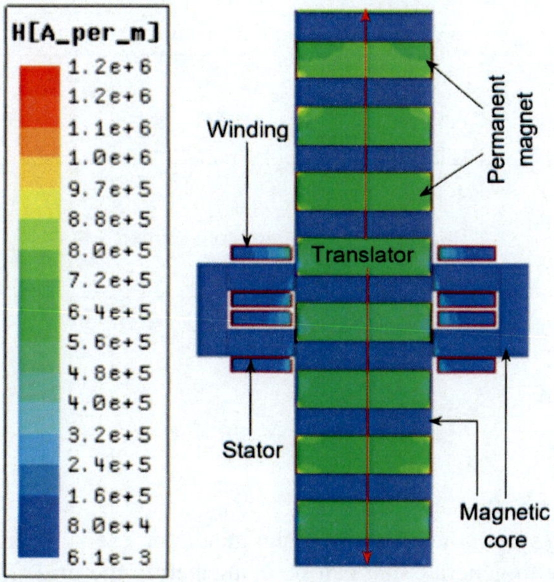

Fig. 23. Field intensity of a proposed linear generator

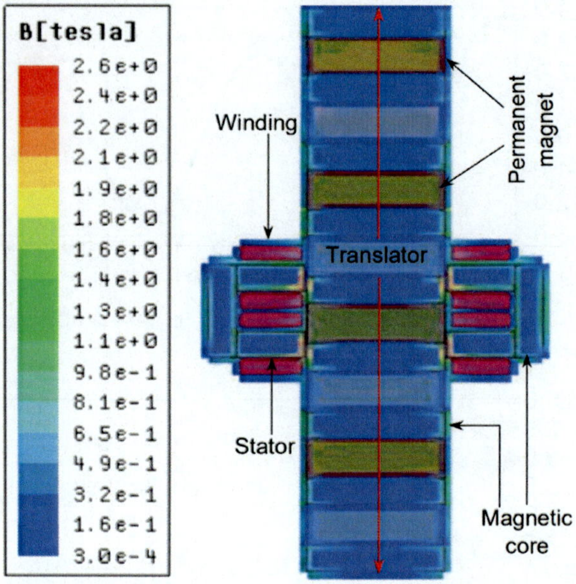

Fig. 24. Flux density of a proposed linear generator

Table 4. Parameters of the proposed generator

Item	Parameter	Value
Translator	Length (mm)	336
	Width (mm)	80
	Thickness of PM pole (mm)	24
	Number of PM	8
	Thickness of translator core (mm)	16
	Number of translator core	9
Stator	Height (mm)	75
	Width (mm)	70
Model depth (mm)		500
Air gap (mm)		2

By increasing size of the proposed linear generator, even higher amount of electrical power is possible to generate that can be connected to the grid. Thus, by utilizing the oceanic wave energy, mitigation of environmental pollution is possible.

7 Conclusion

This paper proposes a linear electrical generator which is small and simple in construction. But it produces up to 5.44 kW electrical power that is quite high compared to its size. Simulation results show the excellent performance of the linear generator by plotting different parameters with respect to time. Magnetic properties of the proposed linear generator are also presented with the scale (rainbow spectrum). As the oceanic wave energy is enormous, and they are still almost untapped. The motivation of this paper is to produce electrical power from the oceanic wave, which is one of the promising RESs. As the proposed linear generator would be driven by oceanic wave energy, it can be a suitable solution for the future power generation. The oceanic wave energy based power plant or wave farm consists of several linear electrical generators. Therefore, the proposed linear generator can be cascaded to increase the power generation capability as required. It would be not only suitable for Bangladesh but also suitable for many countries that have not enough land area and other RESs. It is expected that it can be supportive for the net zero carbon emission target for many countries.

References

1. Power generation: http://119.40.95.168/bpdb/power_generation_unit. Accessed 23 Oct 2022
2. Clément, A.: Wave energy in Europe: current status and perspectives. Renew. Sustain. Energy Rev. **6**(2), 405–431 (2002)
3. Pontes, M.T.: Assessing the European wave energy resource. J. Offshore Mech. Arct. Eng. **120**(4), 226–231 (1998)
4. Molla, S., Farrok, O., Islam, M. R., Xu W.: The novel low reluctance superconducting permanent magnet linear generator for oceanic wave energy extraction, IEEE Trans. Appl. Supercond. **31**(8), Art. ID 5203705 (2021)
5. Liu, C., Rui, D., Zhu, H., Fu, W.: Multi-physical coupling field of a permanent magnet linear synchronous generator for wave energy conversion. IEEE Access **9**, 85738–85747 (2021)
6. Molla, S., Farrok, O., Islam, M. R., Muttaqi, K.M.: Analysis and design of a high performance linear generator with high grade magnetic cores and high temperature superconducting coils for oceanic wave energy conversion, IEEE Trans. Appl. Supercond., **29**(2), Art. ID. 5201105 (2019)
7. Molla, S., Farrok, O. Use of high flux density ferromagnetic cores in linear generators for oceanic wave energy conversion. In: 4th International Conference on Electrical, Communication and Computer Engineering, pp. 7–9, Cox'sbazar, Bangladesh (2019)
8. RE generation mix. http://www.renewableenergy.gov.bd/. Accessed 23 Oct 2022
9. World energy consumption by energy source 2050. https://www.statista.com/statistics/222066/projected-global-energy-consumption-by-source/. Accessed 12 Sep 2022
10. Electricity generation. https://ourworldindata.org/grapher/electricity-generation?tab=chart&country=~BGD. Accessed 12 Sep 2022
11. Bangladesh: CO_2 country profile. https://ourworldindata.org/co2/country/bangladesh. Accessed 15 Oct 2022
12. Karmaker, A.K., Rahman, M.M., Hossain, M.A.M., Ahmed, R.: Exploration and corrective measures of greenhouse gas emission from fossil fuel power stations for Bangladesh. J. Cleaner Prod. **244**, Art. ID 118645 (2020)
13. https://www.ucsusa.org/resources/environmental-impacts-solar-power#references. Accessed 19 Oct 2022

14. Rahman, A., Farrok, O., Haque, M.M.: Environmental impact of renewable energy source based electrical power plants: Solar, wind, hydroelectric, biomass, geothermal, tidal, ocean, and osmotic. Renew. Sustain. Energy Rev. **61**, Art. ID 112279 (2022)
15. Gao, C.K., et al.: Environmental impact analysis of power generation from biomass and wind farms in different locations. Renew. Sustain. Energy Rev. **102**, 307–317 (2019)
16. Hondo, H.: Life cycle GHG emission analysis of power generation systems: Japan. Energy **30**(11), 2042–2056 (2005)
17. Varun, Prakash, R., Bhat, I.K.: Life cycle energy and GHG analysis of hydroelectric power development in India. Int. J. Green Energy **7**(4), 361–375 (2010)
18. Xu, X., Tan, Y., Guishan, Y.: Environmental impact assessments of the Three Gorges Project in China: Issues and interventions. Earth Sci. Rev. **124**, 115–125 (2013)
19. Biomass and the environment: https://www.eia.gov/energyexplained/biomass/biomass-and-the-environment.php. Accessed 11 Oct 2022
20. Kumar, V.S., Anoop, T.R.: Wave energy resource assessment for the Indian shelf seas. Renew. Energy **76**, 212–219 (2015)

The Effect of Air Density in Offshore Wind Power Potential in India

Garlapati Nagababu[1](✉), Ravi Patel[2], and Kantipudi M. V. V. Prasad[3]

[1] School of Technology, Pandit Deendayal Energy University, Gandhinagar, Gujarat, India
garlapatinagu@gmail.com

[2] Renewable Energy, Torrent Power, Surat, Gujarat, India

[3] Symbiosis Institute of Technology, Symbiosis International (Deemed University), Pune, Maharashtra, India

Abstract. In the wind energy sector, the yearly energy production is often calculated using a constant value of air density based on the location's annual average value. As a result, the correction necessary in assessing daily, monthly, or seasonal wind power density (WPD) owing to air density is typically neglected in current literature. Therefore, the present study assessed the WPD due to variations in air density. Indian offshore region is considered as a study area. Further, wind speed, air density and temperature data are taken from the latest ERA5 reanalysis. The findings show that the WPD is higher by 5% as determined by when the air density adjustment is taken into account. Seasonal results show that during summer, WPD has a maximum variation of about 7–8% and a minimum variation of about 2–5.5% during winter.

Keywords: Wind power density · Standard air density · Real air density · ERA5 · India

1 Introduction

Making an effective energy strategy requires understanding offshore wind energy in the future. However, it is still unclear how offshore wind energy will adapt to the rapidly escalating global warming. Harnessing offshore energy help to achieve carbon neutrality. However, the availability of wind resources is susceptible to climate change owing to which it is imperative to investigate if baseload power is harnessed using the installed plants. Due to the fact that wind speeds over the ocean are often more consistent and powerful than those over land regions, offshore wind energy has a major advantage in terms of its capacity to generate more electricity. WPD refers to the quantity of wind energy accessible at a wind turbine location. It is directly proportional to air density and the cube of wind speed. It is independent of a wind turbine, so it is considered as a measure of wind energy potential. Studies focusing on air density and its impact on wind power were conducted in a much smaller number. When air density varied from -5% to $+5\%$, wind energy generation lies between -2.4% and $+2.22\%$, respectively [1]. Zhang et al. [2] presented multivariate and multimodal wind distribution for the

S. Haddout et al. (Eds.): CCORE 2022, SPEES, pp. 23–29, 2023.
https://doi.org/10.1007/978-3-031-26967-7_2

onshore and offshore sites in the USA by considering wind speed, air density and wind direction. Carta et al. [3] considered a variation of density and wind speed in a bivariate probability for Spain. In another study, air density distribution was analysed using a uniform distribution.

In previous wind resource assessments, spatial and temporal variations of air density are hardly considered. Moreover, the air density value is considered a constant (1.225 kg/m^3) throughout space and time. This constant value is applicable only at standard conditions with a temperature of 15 °C and pressure of 1013 hPa. In general, pressure governs the significant spatiotemporal fluctuations of air density. Temporal variability of pressure depends on meteorological conditions and spatial variability of pressure depends on the height. Therefore, air density shows inter- and intra- annual variability as well as long-term trends [4]. Farkas [5] assessed the temporal air density variation at a location in Hungary and observed a variation of 15% from its mean. These variations in air density have a proportionate impact on WPD. Due to air density variation, offshore wind power generation uncertainty over Scotland reaches up to 3% during winter and instantaneous power fluctuates by more than 26% [6]. Over Iberian Peninsula, WPD fluctuates around 8% between winter and summer with and without air density adjustment [7]. A study by Collins et al. [8] observed that on cold and hot days, wind turbines' power output varied by around 10% at the same magnitude of wind speeds. Recent studies focused on Indian offshore region [9–11] have taken the constant air density value.

Presently available research on the impact of air density in wind resource assessment is limited. Moreover, these studies are site-specific, providing only a limited understanding of the function of air density in wind energy evaluation. Comprehensive spatial distribution of air density allows us to assess its impact on wind energy calculation. In the Indian offshore region, the effect of air density variability on wind power has not been evaluated yet. Moreover, air density has spatiotemporal variability, and it depends on elevation, pressure, and meteorological conditions. This will lead to a large deviation in the projected and actual plant load factor of an offshore wind farm. Ultimately this results in huge financial losses for the investors. Therefore, in the present study, a spatiotemporal influence of air density on WPD along the Indian offshore region is evaluated.

2 Data and Methods

2.1 Data

The ERA5 hourly data are collected from European Centre for Medium-Range Weather Forecasting (ECMWF) [12]. ERA5 is the recently developed reanalysis dataset and has wind data from 1979 onwards in different temporal resolutions (hourly, daily, monthly and so on). ECMWF constantly improved the dataset's accuracy and spatial and temporal resolution. In the open literature, various reanalyse datasets (MERRA-1/2, ERA-Interim, NCEP/NCAR and NCEP/CFSR) were used for offshore wind energy applications. However, the main drawback of these reanalysis datasets is they provide wind data at 10 m above mean sea level (msl). Generally, an offshore wind turbine's height is more than 80 m. In order to estimate the performance of the wind turbine's output, wind speed data is required at that height. ERA5 is the only reanalysis data that provide wind speed at 100 m above msl. Moreover, it provides a higher spatial resolution of 31 km and 1-h temporal resolution data. Therefore, in the present study, 19 years (2000 to 2018) of hourly data of the following parameters are retrieved from ECMWF: wind speed at 100 m above msl, temperature and air density at 2 m and 10 m above msl over Indian offshore region (24°N 65°E to 4°N 90°E).

2.2 Methodology

The retrieved time series data of temperature data from ECMWF is at 2 m above msl. This temperature data were extrapolated to hub height (100 m) by Eq. (1)

$$T_2 = T_1 + L(H_2 - H_1) \tag{1}$$

where T_1 and T_2 are the temperature at the height of H_1 and H_2 respectively. L is the temperature lapse rate, –0.0065 K/m [13]. The air density at 100 m was estimated by Eq. (2) [14]

$$\rho(H_{hub}) = \rho\left(H_{ref}\right)\left(\frac{T(H_{hub})}{T\left(H_{ref}\right)}\right)^{\left(\frac{g}{LR}-1\right)} \tag{2}$$

where $\rho(H_{hub})$ and $\rho\left(H_{ref}\right)$ are the air densities at hub height (100 m) and reference height (10 m). R is the ideal gas constant, and g is the acceleration of gravity (9.8 m/s^2).

WPD is the amount of kinetic wind power per unit swept area. It is the function of air density (ρ) and wind speed (U) in the time series of N cases i = 1,2,..n. It can be estimated by Eq. (3).

$$WPD = \frac{1}{2}\sum_{i=1}^{N} \rho_i\, U_i^3 \tag{3}$$

Generally, standard air density ($\rho_0 = 1.225$ kg/m^3)was used in the WPD calculations. Air density depends on various factors like temperature, humidity and pressure. These values vary with respect to time and place. Therefore, variation in air density can be considered in WPD calculations. The percentage relative error (ΔWPD) between the WPD with constant air density (WPD_0) and time-series air density (WPD) is calculated by Eq. (4).

$$\Delta\, WPD\,(\%) = \frac{WPD_0 - WPD}{WPD} \times 100 = \left(\frac{\sum_{i=1}^{N} \rho_i U_i^3}{\rho_0 \sum_{i=1}^{N} U_i^3} - 1\right) \times 100 \tag{4}$$

3 Result and Discussion

Spatial distribution of WPD based on constant air density value and time-series air density is presented in Fig. 1. From Fig. 1, it has been noticed that WPD across study area is slightly higher with constant air density than time-series air density. Although highest and lowest energy regions for both maps are looks similar. However, magnitude of WPD with constant air density is 5–13 W/m^2 higher than time-series air density. The southern tip of study area has the strongest winds, with mean WPDs ranging from 400 to

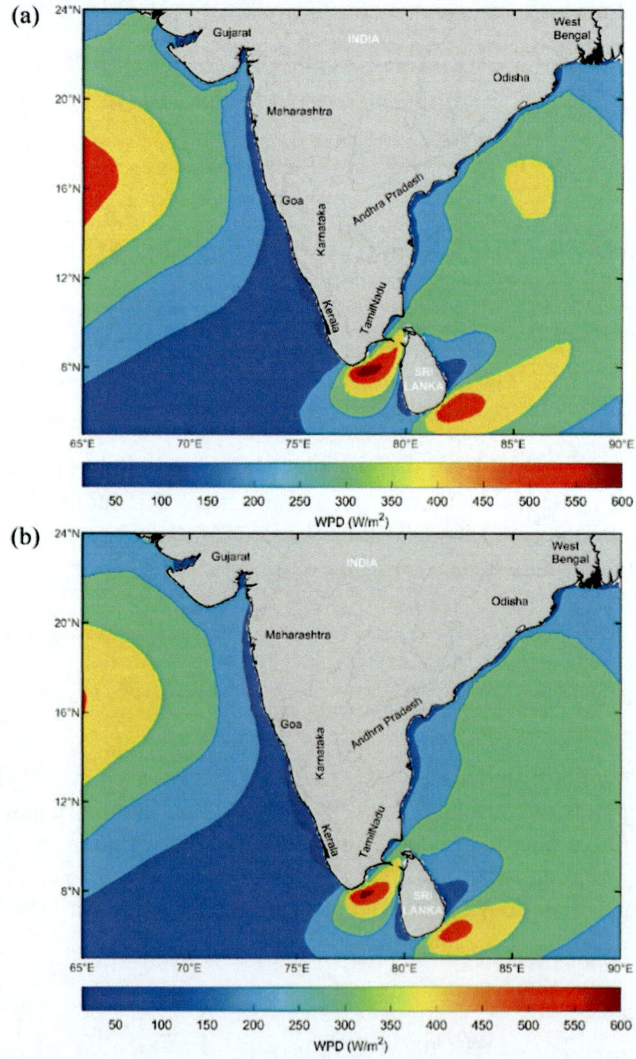

Fig. 1. Spatial distribution of WPD along Indian offshore region by using (a) constant air density value (1.225 kg/m^3) (b) time-series value of air density (period: 2000–2018)

600 W/m^2. Subsequently, offshore region of Gujarat, Odisha and Andhra Pradesh experience mean WPD of 300–450 W/m^2. Southern part of western coast (Goa, Karnataka and Kerala) have low wind potential with a mean WPD of 50–100 W/m^2.

Using Eq. (4), the percentage change in WPD caused by time-series air density is determined for the entire study region and presented in Fig. 2. From Fig. 2, it can be observed that the percentage relative error of mean WPD is 5 to 8% between the constant and time-series air density assumption. This could be due to the relatively higher temperature of the region, which resulted in lower (1.15- 1.16 kg/m^3) air density than the constant (1.225 kg/m^3). As a result, the percentage relative error of mean WPD is positive. Higher fluctuations of WPD (6–7.5%) due to air density are observed along the eastern coast compared to the southern and western coasts of India. Along the southernmost tip and a southern part of the western coast, WPD changes slightly more than 5 to 5.5% are observed.

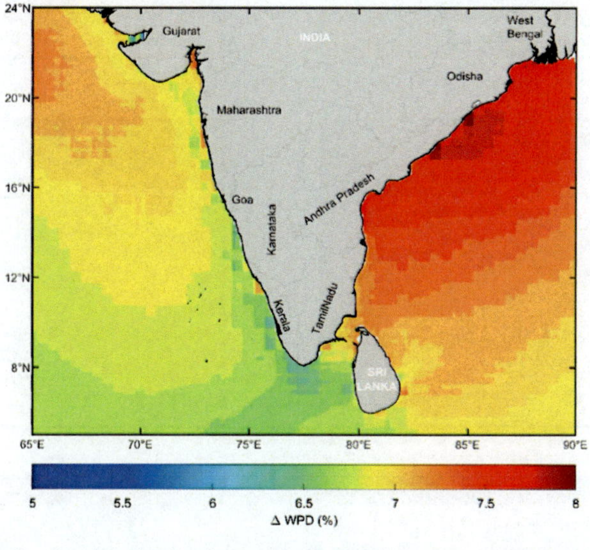

Fig. 2. Spatial distribution map of ΔWPD in terms of percentage

India's climate includes a wide variety of weather patterns over a huge geographic area and diverse topography. Additionally, the Bay of Bengal and the Arabian Sea significantly impact the climate. The Bay of Bengal is significantly warmer than the Arabian Sea, and many more storms develop over the bay, even though both are situated in the same latitude area and get the same amount of solar energy from the Sun. Therefore, seasonal fluctuations in air density may be anticipated to be responsible for the WPD changes. While estimating seasonal WPD, four seasons have been considered as per Indian climate conditions such as summer, monsoon, autumn and winter. Summer includes April, May and June. Monsoon includes July, August, and September. Autumn includes October and November. Winter includes December, January, February and March. Figure 3 illustrates the spatial maps of the change in WPD with the season to better understand how it is affected by seasonal variations in air density. Summer has the lowest air density compared to the remaining seasons due to higher temperatures prevailing during this season. As a result, this season had the greatest variation in WPD (7–8%). In contrast, due to cold weather, the least fluctuations (2–5.5%) are observed in winter. Assuming the standard air density would not have been able to account for these fluctuations in WPD with season.

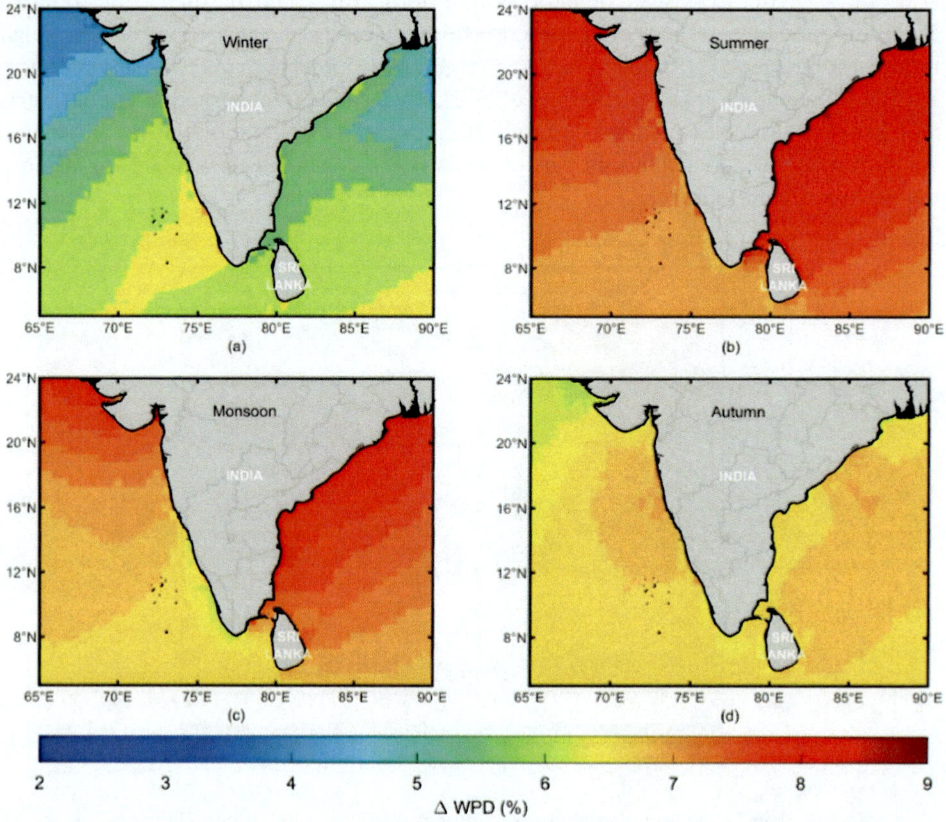

Fig. 3. Spatial distribution maps of seasonal ΔWPD in percentage (a) Winter (b) Summer (c) Monsoon (d) Autumn

4 Conclusion

The present study assessed the WPD due to variations in air density along the Indian offshore region. Meteorological parameters such as wind speed, air density, and temperature data are taken from the latest ERA5 reanalysis produced by ECMWF. The obtained data are at a spatial resolution of 0.1 and hourly temporal resolution from 2000 to 2018. By using the data, seasonal and annual deviation of WPD due to air density was analysed. The spatial distribution of air density values is lower (1.15–1.16 kg/m^3)

than the constant value (1.225 kg/m^3). The results demonstrate that when the air density adjustment is included, the WPD is 5% greater. According to seasonal data, WPD varies the most during the summer by around 7–8% and the least during the winter by about 2–5.5%. Significant spatial and seasonal fluctuations in WPD have been observed along the Indian offshore region, implying the necessity for a regionalized strategy. The methodology presented in this study will help to accurately estimate wind resources for offshore wind farm development in India. A more accurate estimation of wind potential will bring the confidence of investors for investing in this field of energy. Further, the findings of this study will assist in precise micro-siting of wind turbine locations.

References

1. Villena-Ruiz, R., Ramirez, F.J., Honrubia-Escribano, A., Gómez-Lázaro, E.: A techno-economic analysis of a real wind farm repowering experience: the Malpica case. Energy Convers Manag. **172**, 182–199 (2018). https://doi.org/10.1016/J.ENCONMAN.2018.07.024
2. Zhang, J., Chowdhury, S., Messac, A., Castillo, L.: A multivariate and multimodal wind distribution model. Renew Energy **51**, 436–447 (2013). https://doi.org/10.1016/J.RENENE.2012.09.026
3. Carta, J.A., Mentado, D.: A continuous bivariate model for wind power density and wind turbine energy output estimations. Energy Convers Manag **48**, 420–432 (2007). https://doi.org/10.1016/J.ENCONMAN.2006.06.019
4. Jung, C., Schindler, D.: The role of air density in wind energy assessment – a case study from Germany. Energy **171**, 385–392 (2019). https://doi.org/10.1016/j.energy.2019.01.041
5. Farkas, Z.: Considering Air Density in Wind Power Production (2011)
6. Ulazia, A., Nafarrate, A., Ibarra-Berastegi, G., Sáenz, J., Carreno-Madinabeitia, S.: The consequences of air density variations over northeastern Scotland for offshore wind energy potential. Energies **12** (2019). https://doi.org/10.3390/en12132635
7. Ulazia, A., Ibarra-Berastegi, G., Sáenz, J., Carreno-Madinabeitia, S., González-Rojí, S.J.: Seasonal correction of offshore wind energy potential due to air density: case of the Iberian Peninsula. Sustain **11**, 1–22 (2019). https://doi.org/10.3390/su11133648
8. Collins, J., Parkes, J., Tindal, A.: Short term forecasting for utility-scale wind farms — the power model challenge. Wind Eng. **33**, 247–257 (2009). https://doi.org/10.1260/0309-524X.33.3.247
9. Nagababu, G., Surendra Singh, K., Vimal, S., Ranajit, B.: Evaluation of offshore wind power potential in the western coast of India: a preliminary study. Curr. Sci. **112**, 62–67 (2017). https://doi.org/10.18520/cs/v112/i01/62-67
10. Nagababu, G., Kachhwaha, S.S., Savsani, V.: Estimation of technical and economic potential of offshore wind along the coast of India. Energy **138**, 79–91 (2017). https://doi.org/10.1016/j.energy.2017.07.032
11. Nagababu, G., Naidu, N., Kachhwaha, S., Savsani, V:. Feasibility study for offshore wind power development in India based on bathymetry and reanalysis data. Energy Sources Part A Recover. Util. Environ. Eff. 2016. https://doi.org/10.1080/15567036.2016.1233303
12. Hersbach, H., Bell, B., Berrisford, P., Hirahara, S., Horányi, A., Muñoz-Sabater, J., et al.: The ERA5 global reanalysis. Q. J. R. Meteorol. Soc. **146**, 1999–2049 (2020). https://doi.org/10.1002/QJ.3803
13. United States Committee on Extension to the Standard Atmosphere. U.S. Standard Atmospher, 1976 (1976)
14. Patel, R.P., Nagababu, G., Singh, S., Kumar, V.V.A.: A revised offshore wind resource assessment and site selection along the Indian coast using ERA5 near-hub-height wind products. Ocean Eng. **254**, 11134 (2022). https://doi.org/10.1016/j.oceaneng.2022.111341

Influence of Vertical Plates on the Pitching Motion of a SPAR Wind Floater in Waves

Fuad Mahfud Assidiq(✉), Daeng Paroka, Habibi, Hidayatullah, and Muhammad FajarFitra Ramadan

Department of Ocean Engineering, Universitas Hasanuddin, Bontomarannu, Gowa, South Sulawesi, Indonesia
{assidiqfm,habibi}@unhas.ac.id, dparoka@eng.unhas.ac.id, hidayatullah20d@student.unhas.ac.id

Abstract. One of the concerns about using a heave plate on a SPAR floater is that the pitch response will be inadequately minimized. This weakness will result in excessive misalignment conditions in the floater wind. The vortex ring state phenomenon is a more complex effect and the most common potential issue. This issue may result in over-performance of wind turbine system components. A time domain based experimental test was developed in this study to investigate the extent to which the number and configuration of vertical plates influence pitch response, which is useful for improving wind energy extraction performance. Because the vertical plate configuration is an important key parameter, the following arrangements are proposed to be investigated: solid plates, three-perforated plates, and six-perforated plates. 3VP-S, 3VP-3H, 3VP-6H, 4VP-S, 4VP-3H, 4VP-6H, 5VP-S, 5VP-3H, and 5VP-6H are the modified SPARs. The ballasting area is where the modified vertical plate of the base SPAR is located. In 120 regular wave cycles, the effects of the proposed model's pitch response, pitch reduction percentage, wave steepness, natural period, and viscous damping were evaluated. The wave tank was used to test ten different scenarios. The 4VP-6H, 5VP-3H, and 4VP-S models were identified to provide better estimates than other modified models. The perforated plate is confirmed to be more effective at reducing pitch motion than the solid plate and base SPAR. The increase in pitch response is consistent with the increase in applied wave parameters. The model was also initiated to be in dominant stand-by mode for all wave parameters. Because of the reduced volume caused by the SPAR body modification, the percentage reduction in pitch motion increases with decreasing wave loads. When compared to the base SPAR, the proposed model's pitch reduction percentage ranges from 19 to 56%. In all scenarios, the wave steepness relationship is inversely proportional to the pitch reduction percentage. Furthermore, none of the proposed models are closer to the natural frequency than the base SPAR, resulting in the lowest resonance. The perforated plate also dampens more effectively than the base SPAR. This is primarily due to the geometry of the vertical plate and its configuration layout. Finally, as the most recent conceptual design in wind floater SPAR, the 4VP-6H model is deeply recommended.

Keywords: Pitch responses · SPAR floater wind · Vertical plates · Wave parameters

S. Haddout et al. (Eds.): CCORE 2022, SPEES, pp. 30–42, 2023.
https://doi.org/10.1007/978-3-031-26967-7_3

1 Introduction

The SPAR floater is essentially one of the fundamental alternatives for a substructure for wind turbine floaters. When its relative location is between the center of gravity and the center of buoyancy, SPAR floater stability manifests unique properties [1]. The buoyancy force generated by the spar platform supports the entire offshore wind turbine, and the mooring rope tension maintains the spar-type flying substructure's station position. The mooring system has the greatest influence on surge, sway, and yaw, whereas the floating platform's characteristics have the major impact on heave, roll, and pitch [2]. Comparatively speaking, SPAR floaters are relatively inexpensive. This is due to the small submerged volume, which produces a negligible wave force. The SPAR floater's natural period is also more advantageous in these conditions. On the other hand, the floater's minimal shape could lead to excessive fatigue in the area around its tower. Additionally, it requires a depth of more than 150 m [3, 4].

In the development of wind turbine floater technology, heave plates are one of the methods being manipulated to reduce heave motion [5, 6]. A heave plate is a plate that is attached to the bottom of a column with the intention of increasing the system's additional mass and damping and, as a result, lowering the vertical response in FOWT waves. As can be seen in [7], the wind floater type semi-submersible OC4 presents this device to improve the hydrodynamic behavior. Heave plate structures have been studied for single column platforms with a focus on the hydrodynamic effects, particularly on the reduction of heave motion. For instance, numerical analysis was carried out to optimize the heave plate geometry in [8, 9]. A practical illustration of the investigation of the hydrodynamic properties of heave plates is provided by [10]. Heave plate structures can be used to lessen heave motions and can also be applied to multi-column systems, according to both studies. However, according to the previous explanation, the dominant heave plate only scales heave motion down. Another reason is that the SPAR floater has a large draft, which allows for a relatively small heave motion [11]. This is why heave plate technology is incompatible with SPAR floaters. In SPAR floaters, pitch motion is more dominant, necessitating the use of a pitch control system [12]. There has been little direct investigation into vertical plate technology. SPAR floater experiments with various vertical plate configurations were carried out by Ishida and Imai [13]. It was observed that the solid vertical plate increased damping more effectively than the heave plate. The leading reason for this is that the natural period of the solid vertical plate is shorter due to the lower mass in the ballasting zone when compared to the base SPAR floater.

Motivated by the preceding discussion, the purpose of this study is to investigate the effect of vertical plate configuration on pitch motion, which is proper for improving wind energy extraction performance based on experimental observations. Furthermore, the discussion includes a performance comparison of the base SPAR floater and the modified SPAR model, with an impressiveness on the pitch response of various cases, the pitch response reduction of the proposed model, the effects of wave steepness, pitch natural period, and viscous damping of the proposed model.

2 Concept of Vertical Plate

Heave plate technology has become increasingly popular in wind floater in recent years. When compared to the base model, this approved combination produces acceptable station keeping capability [14, 15]. The heave plate, a horizontal plate, is typically installed on the keel or main body. Most previous studies found that heave plates could only contribute viscous damping to prevent heave motion. Pitch motion, on the other hand, can propagate significant turbulence during the wind energy extraction process [16–18]. Therefore, plate technology is required to reduce pitch motion.

The concept presented in this study is in the arrangement of a vertical plate. The vertical plate concept should provide viscous damping to depress pitch motion. According to references [2, 19], the SPAR base model is as follows. The vertical plate arrangement is directly integrated into the main body to replace the ballasting zone of the base SPAR. Because the vertical plate configuration is an important key parameter, the proposed arrangements to be investigated are as shown in Fig. 1: the number of vertical plates, solid plates, three-perforated plates, and six-perforated plates. To make data collection easier for all models, names like 3VP-6H are used, which means that the SPAR uses three vertical plates with six holes. Figure 2 and Table 1 summarizes the prototype and scale model's key parameters. Given the limited dimensions of the wave tank used, a geometric model scale of 1:115 was purposed in this study. Table 2 also categorizes the pitch response parameters practiced to the SPAR.

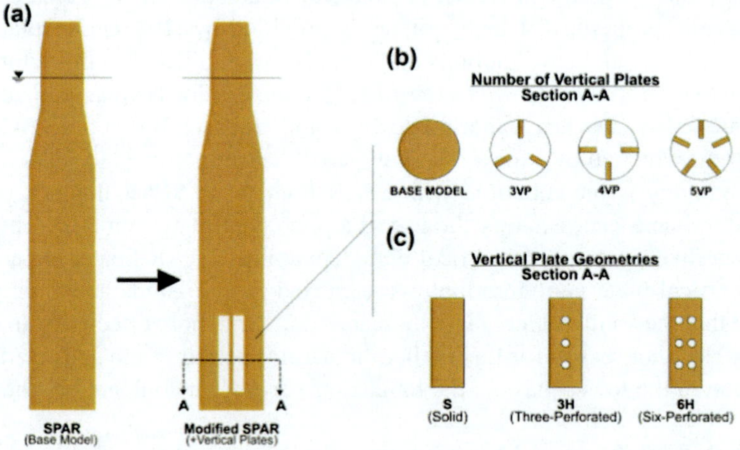

Fig. 1. Schematic of a 6 MW SPAR floater wind model; (a) modifying the base model to a novel SPAR design with vertical plates, (b) number of vertical plates: 3VP, 4VP, and 5VP, (c) geometry of vertical plates: 3VP, 4VP,and 5VP based on a view of cross-section A-A

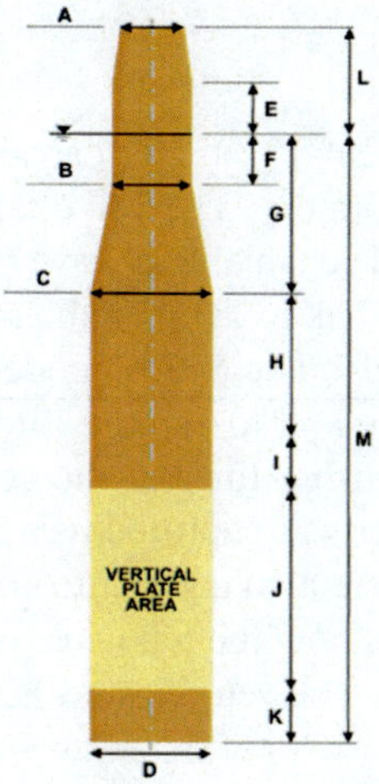

Fig. 2. Sizing in SPAR model design

Table 1. SPAR model dimensions, scale 1:115 applied according to wave tank

Codes	Values	Units
A	0.07	m
B	0.08	
C	0.13	
D	0.13	
E	0.05	
F	0.05	
G	0.17	
H	0.25	
I	0.04	
J	0.15	
K	0.04	
L	0.11	
M	0.66	

Table 2. Various pitching mode criteria on SPAR floater winds

Conditions	Criteria
Operating	$0 < \text{Pitch} \leq 0.7$
Survival	$0.7 < \text{Pitch} \leq 2$
Stand by	$2 < \text{Pitch} \leq 6$
Damaged	$6 < \text{Pitch} \leq 18$

3 Methodology

This wave tank-based experimental test was carried out at Universitas Hasanuddin's Ocean and Environmental Laboratory. The wave tank measures 24 m long, 1 m wide, and has a water depth of 0.9 m. The tank has a wave maker at the end. A wave absorber is installed at the other end of the tank to dampen the reflected waves. Figure 3 depicts the experimental setup schematically. The SPAR model is positioned in the tank's center. To keep the SPAR from drifting and to capture pitch motion, the model is tethered in tension by two thin rubber mooring lines at the center of gravity. The time domain-based pitch motion of the model was captured using a gyroscope tracking system. The capabilities of the wave tank were also taken into consideration when choosing the test wave parameters. Table 3 shows how the ten wave parameters describe the zero-degree directional wave incident angle. The setup lasted 120 s.

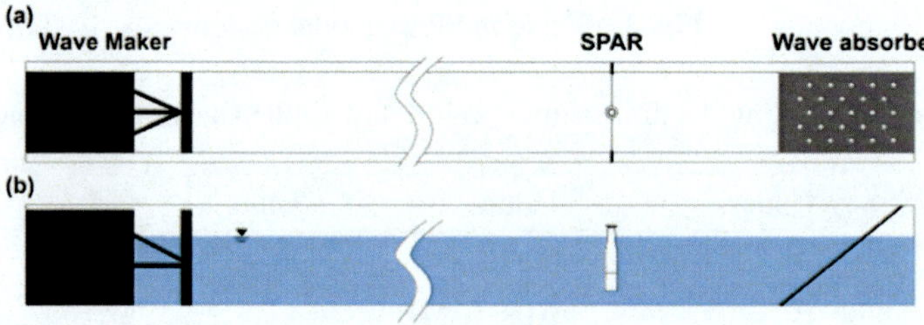

Fig. 3. SPAR model setup on the wave tank, (a) top view, and (b) side view

Table 3. Wave parameter scenarios on the SPAR model, scale 1:115 applied according to wave tank

Cases	H (m)	T (s)
1	0.0087	1
2	0.0174	1
3	0.0261	1
4	0.0174	1.2
5	0.0261	1.2
6	0.0348	1.2
7	0.0522	1.2
8	0.0261	1.4
9	0.0348	1.4
10	0.0522	1.4

4 Results and Discussions

The Effect of the Number of Vertical Solid Plates on the Spar Pitch Motion
Figure 4 illustrates how the SPAR's pitch response is dependent on the number of vertical solid plates. The statistically based response, which includes the maximum, minimum, mean, and standard deviation, plots the variations in pitch characteristics. The most effective models depend on the wave parameters are 4VP-S, 5VP-S, 3VP-S, and the base model, respectively. The pitch motion increases along with the wave height. The same observation holds true as the wave period increases. However, the pitch motion is not always affected by adding more vertical solid plates. The 4VP-S model clearly outperforms both the 3VP-S and 5VP-S models in terms of explaining this effect. The direction of the incident wave and the vertical plate configuration itself have a dominant impact on this. The vertical solid plate configuration modifies the direction of flow separation to specifically alter the drag coefficient. Additionally, the impact of the wave incident direction creates an angle of attack between each vertical plate. The difference in angle of attack causes a sizable variation in pitch motion. As a result, it is advised that the 4VP-S model be applied to best mitigate the pitch response.

Pitch Motions of Three-Perforated Vertical Plates Compared to the Base Model
Figure 5 shows the comparison between the three-perforated vertical plate and the base model. Similar to the previous sub-section, a statistically based approach was also used. The base model, 5VP-3H, 4VP-3H, and 3VP-3H, in order of feasibility, are carefully sorted models. The findings indicate that the pitch motion increases as the wave parameter increases. Another distinctive feature is that the pitch motion decreases as the number of vertical plates increases. It is obvious that the three-slotted scheme can effectively reduce pitch characteristics. The same finding holds true for the impact of the vertical plate configuration itself and the direction of the incident wave on SPAR. Consequently, it is advised to decide the 5VP-3H model to lessen the pitch motion caused by the three-perforated vertical plates.

Pitch Motions of Six-Perforated Vertical Plates Compared to the Base Mode
Figure 6 depicts the impact of pitch response on the six-slotted vertical plate in comparison to the base model. Pitch response increases along with the wave parameters, as was previously explained. The pitch motion can still be reduced, but to a lesser extent, due to the effect of the number of vertical plates. Comparing the 4VP-6H model to the 3VP-6H and 5VP-6H models, it is clear that the 4VP-6H model is the most effective one. The six-perforated plate layout is the main element that significantly contributes. It results in smaller pitch responses for all wave parameters when the three-perforated plate scheme is taken into account. In light of this, the SPAR models can produce the smallest pitch motion are the base model, 4VP-6H, 3VP-6H, and 5VP-6H, respectively.

Hydrodynamic Performance of the Proposed SPAR Models
The proposed model was chosen to investigate further hydrodynamic performance after determining the smallest pitch motion from a comparison of several SPAR models with consideration of the vertical plate review, in this case the 4VP-S, 4VP-6H, and 5VP-3H

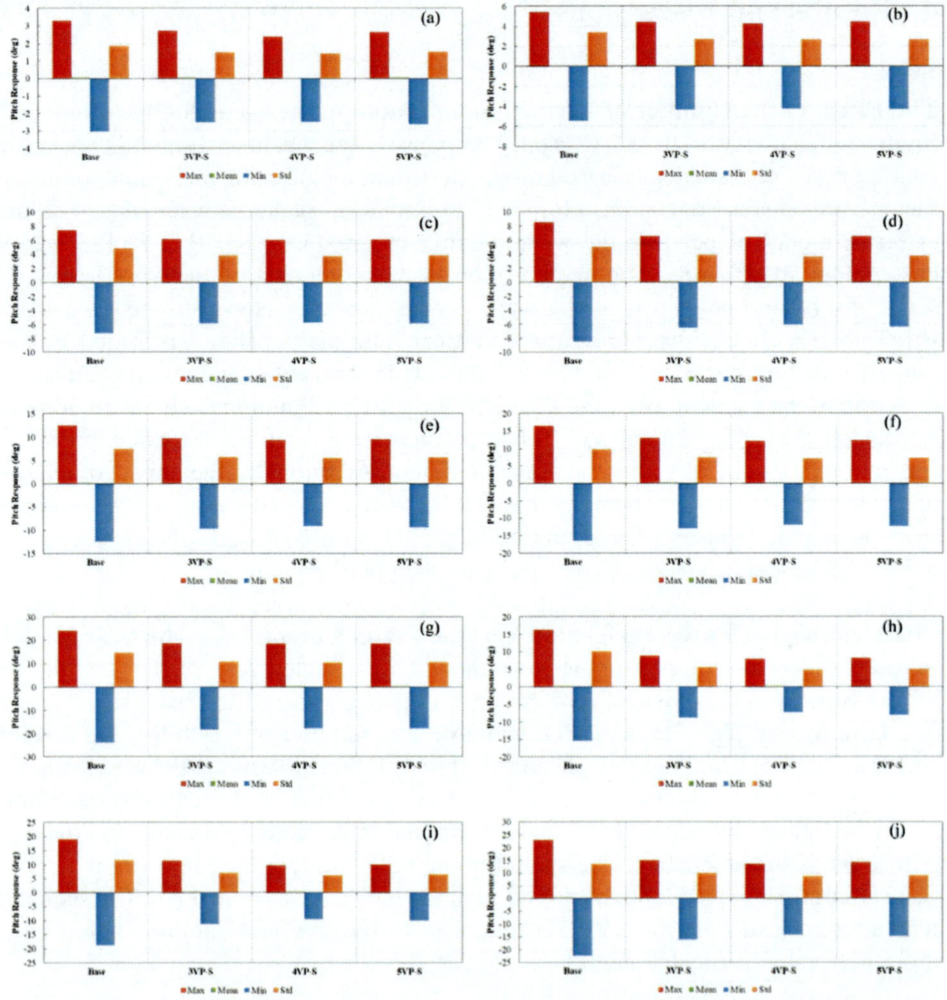

Fig. 4. Pitch responses for SPAR models with different number of vertical plates

models. The following descriptions are given for the hydrodynamic aspects of pitch natural period prediction, non-dimensional damping, wave steepness effect, and the percentage difference of pitch motion.

The first thing is that Fig. 7 lists all of the proposed models along with the pitch reduction percentage. According to previous references in Table 2, the various proposed models are primarily in the stand-by condition, which accounts for 60% of all applied cases. Additionally, a few of the other models are in damaged or survival conditions, with values of 27.5, and 12.5, respectively. In order to decrease the restoring moment, further mooring system planning is therefore required. With regard to the percentage of motion results, it is well known that the percentage of reduced pitch response rises sharply with increasing wave period. The tendency of increasing wavelength is inversely correlated with pitch motion and depends on wave period. On the other hand, the percentage

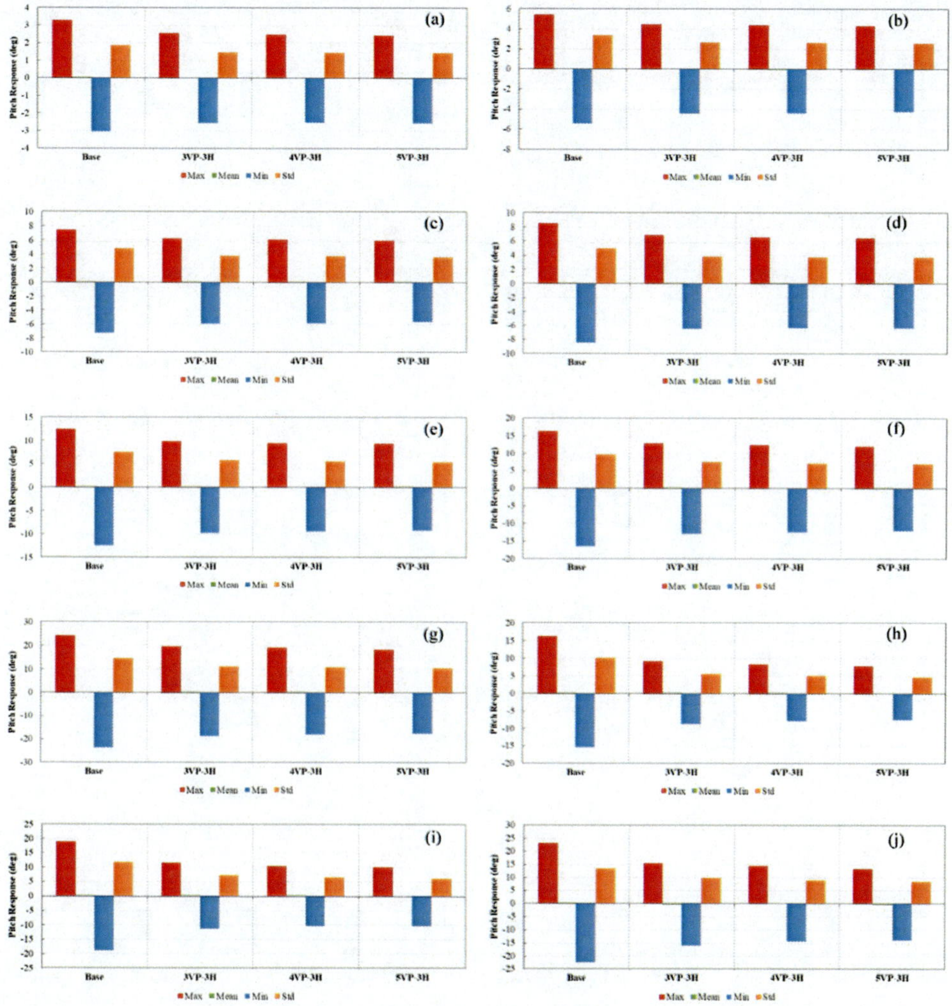

Fig. 5. Pitch responses for SPAR models with three-perforated plates at different configurations.

reduction of pitch motion decreases as the height of the incident wave increases. When passing through the vertical plate configuration, this results in the wave load breaking up. In the end, the proposed models 4VP-6H, 4VP-S, and 5VP-3H are each capable of achieving the best pitch motion reduction percentage.

Second, Fig. 8 depicts the effect of the incident wave steepness H/gT^2 on the pitch response reduction based on the various vertical plate of the SPAR. It is significant to remember that this event was influenced by the incident wave height and incident wave period. The best SPAR models that can achieve the greatest percentage reduction are, in descending order, 4VP-6H, 4VP-S, and 5VP-3H. Pitch motion reduction is less likely as wave steepness rises. As seen in the 4VP-S and 5VP-3H models, the relationship between the incoming wave steepness H/gT^2 and pitch response reduction is linearly decreasing, with a small gradient. Because of the small gradient, the motion reduction

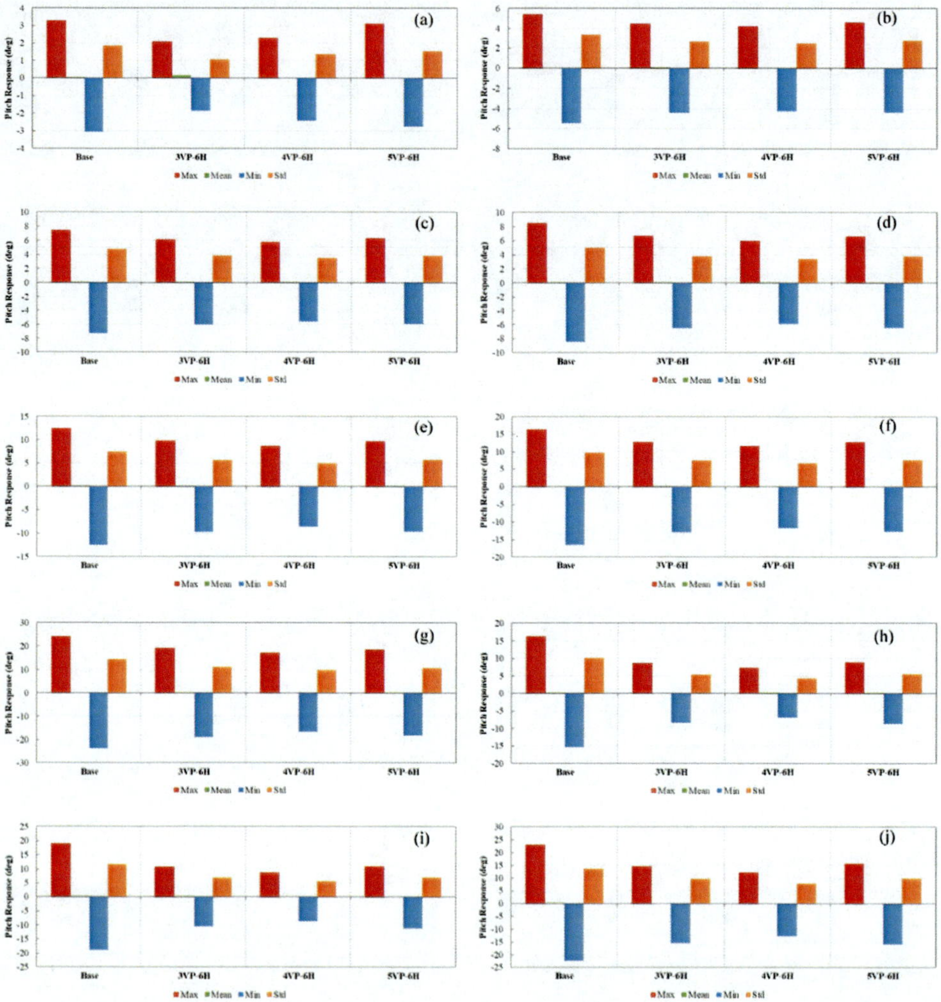

Fig. 6. Pitch responses for SPAR models with six-perforated plates at different configurations

response is not very significant. The gradient of the 4VP-S model is marginally superior to that of the 5VP-3H model when the wave steepness is small. However, as the wave steepness increases, the gradient of the 5VP-3H model slightly outperforms that of the 4VP-S model. The gradient difference for the 4VP-6H model is quite substantial in comparison to the other models. The vertical plate geometry and configuration layout are the main factors affecting this. Accordingly, the effectiveness of the vertical plate variations to reduce SPAR instability events is greatly influenced by the small incident wave steepness in each of the proposed models.

Third, Fig. 9 shows how the non-dimensional damping K and natural period T_n have an impact on the suggested model. Non-dimensional damping is not always directly proportional to the volume immersed in water when the pitch decay test is performed. This is so that the viscous damping can be instantly accommodated by the perforated

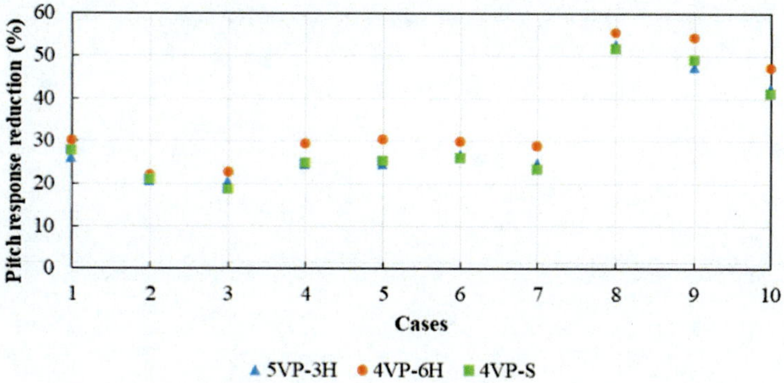

Fig. 7. Pitch response reductions for proposed models in all cases

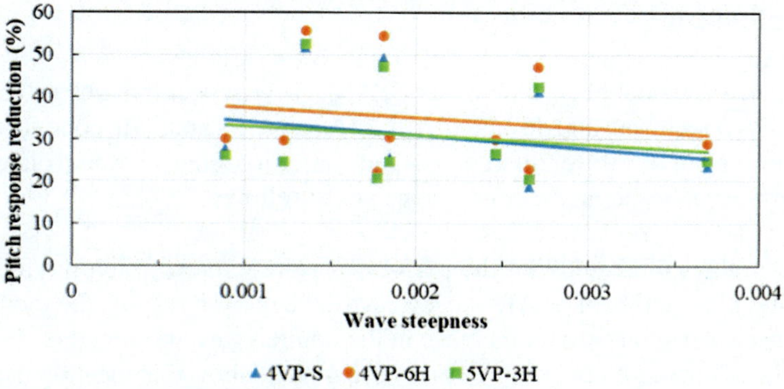

Fig. 8. Pitch response reductions for proposed models at different wave steepness

plate. It demonstrates that the models are capable of producing good damping in the order of 4VP-6H, 5VP-3H, 4VP-S, and base model, respectively. The natural period, on the other hand, is directly proportional to the volume immersed in water. Certain wave period characteristics, ranging from 1 to 1.4 s, have the potential to trigger resonance events. The models with the lowest resonance are 4VP-6H, 4VP-S, 5VP-3H, and the base model, respectively. It is important to note that the effects of the total mass and the stiffness of the SPAR are the most important factors in these events. As a result, the 4VP-6H model was the most efficient damper and significantly decreased the frequency of vibration.

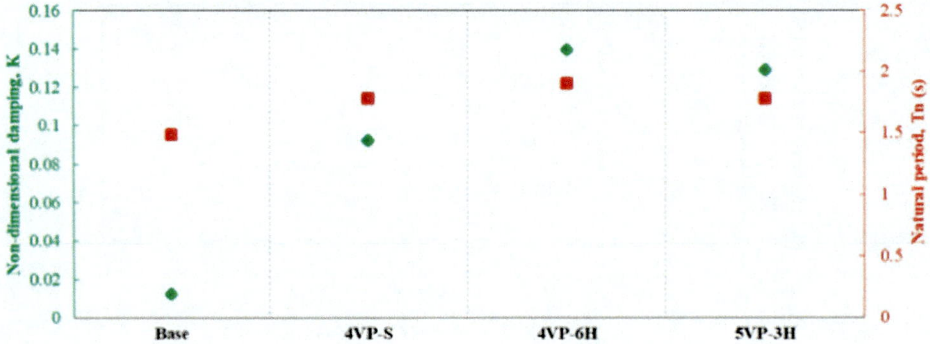

Fig. 9. Non-dimensional damping and pitch natural period of proposed models against base model

5 Conclusions

As a comparison model of common base SPAR, a new conceptual design called vertical plates on SPAR is chosen and discussed in this research. Its goal is to conduct experiments to determine how much different numbers and configurations of vertical plates contribute to pitch response. Some of the conclusions are as follows:

1. Early findings indicated that the perforated plate reduced pitch motion better than the solid plate and base SPAR. According to the motion results, the increase in pitch response corresponds to an increase in the applied wave parameters.
2. For all wave parameters, the three proposed models show that the model is in stand-by mode. The minimum and maximum reduction percentages, for example, in comparison to the base SPAR are 19% and 56%, respectively. The 4VP-6H model has the best motion reduction percentage phenomenon compared to the other two proposed models.
3. At different wave slopes, the three proposed models show varying percentages of motion reduction. The incident wave height and incident wave period dominate the wave slope problem. It means that as the wave slope increases, the pitch motion reduction percentage decreases in the proposed models.
4. Since the three proposed models move in pitch mode, the total mass and stiffness of the SPAR structure determine the natural period and nondimensional damping. All proposed models are more distant from the natural frequency than the base SPAR, resulting in the lowest resonance. The perforated plate's damper effect is the most effective in modifying the overall SPAR structure. Finally, the 4VP-6H model is strongly recommended for use in the most recent conceptual design of a wind floater SPAR.

6 Recommendations

Because this vertical plate study only investigates the SPAR response in terms of pitch motion, other degrees of freedom must be investigated for the proposed vertical plate to perform well. The variation of incident wave direction is also the most important factor to investigate in greater depth for various vertical plate configurations. The offset effect on the SPAR highlights the importance of additional mooring system planning. This should be considered in subsequent studies of this research.

Acknowledgment. The Institute for Research and Community Service (LPPM), Universitas Hasanuddin, Indonesia, provided funding for this research under the Novice Lecturer Research program, also known as PDPU, with Grant No. 1477/UN4.22/PT.01.03/2022.

References

1. Uzunoglu, E., Karmakar, D., Guedes Soares, C.: Floating offshore wind platforms. In: Castro-Santos, L., Diaz-Casas, V. (eds.) Floating Offshore Wind Farms. Green Energy and Technology, pp. 53–77. Springer, Cham (2016). https://doi.org/10.1007/978-3-319-27972-5_4
2. Meng, L., He, Y., Zhou, T., Zhao, Y., Liu, Y.: Research on dynamic response characteristics of 6MW spar-type floating offshore wind turbine. J. Shanghai Jiaotong Univ, (Science) **23**(4), 505–514 (2018). https://doi.org/10.1007/s12204-018-1972-3
3. Chen, J., Kim, M.H.: Review of recent offshore wind turbine research and optimization methodologies in their design. J. Mar. Sci. Eng. **10**, 28 (2022)
4. Bashetty, S., Ozcelik, S.: Review on dynamics of offshore floating wind turbine platforms. Energies **14**(19), 6026 (2021)
5. Amin, I., Dai, S., Day, S., Oterkus, S., Oterkus, E.: Experimental investigation on the influence of interceptor plate on the motion performance of a cylindrical FPSO, Ocean Engineering **243** (2022)
6. Sathia, R., Vijayalakshmi, R.: Effect of keel plate on the performance of fpso suitable for arctic ice environment. Ships Offshore Struct. **15**, 1744–5302 (2019)
7. Robertson, A., Jonkman, J., Masciola, M., Goupee, A., Coulling, A., Luan, C.: Definition of the semisubmersible floating system for phase II of OC4. Technical Report No. NREL/TP-5000-60601, National Renewable Energy Lab. (NREL) (2014)
8. Thiagarajan, K.P., Ran, A.Z., Tao, L., Datta, I., and Halkyard, J.E.: Influence of heave plate geometry on the heave response of classics spars. In: Proceedings of the ASME 2002 21th International Conference on Ocean, Offshore and Arctic Engineering, OMAE 2019–28350, Oslo, Norway (2002)
9. Tao, L., Cai, S.: Heave motion suppression of a Spar with a heave plate. Ocean Eng. **31**, 669–692 (2004)
10. Li, J., Liu, S., Zhao, M., Teng, B.: Experimental investigation of the hydrodynamic characteristics of heave plates using forced oscillation. Ocean Eng. **66**, 82–91 (2013)
11. Henderson, A., Collu, M., Masciola, M.: Overview of floating offshore wind technologies. In: Cruz, J., Atcheson, M. (eds.) Floating Offshore Wind Energy. Green Energy and Technology, Chapter 3, pp. 87–123. Springer, Cham (2016). https://doi.org/10.1007/978-3-319-29398-1_3
12. Goupee, A.J., Koo, B.J., Kimball, R.W., Lambrakos, K.F., Dagher, H.J.: Experimental comparison of three floating wind turbine concepts. J Offshore Mech Arctic Eng **136**, 1–9 (2014)
13. Ishida, S., Imai, Y.: New spar design for floating offshore wind turbine with damping plates. In: 38th International Conference on Ocean, Offshore and Arctic Engineering. Vol. 9, Glasgow, Scotland, UK (2109)
14. Zhao, Ch., Cao, F., Shi, H.: Optimisation of heaving buoy wave energy converter using a combined numerical model. Appl. Ocean Res. **102**, 102208 (2020)

15. Rao, J.M., Nallayarasu, S., Bhattacharyya, S.K.: Numerical and experimental studies of heave damping and added mass of spar with heave plates using forced oscillation. Appl. Ocean Res. **111**, 102667 (2021)
16. Sebastian, T., Lackner, M.A.: Analysis of the induction and wake evolution of an offshore floating wind turbine. Energies **5**, 968–1000 (2012)
17. Sebastian, T., Lackner, M.A.: Characterization of the unsteady aerodynamics of offshore floating wind turbines. Wind Energy **16**, 339–352 (2012)
18. Lamei, A., Hayatdavoodi, M.: On motion analysis and elastic response of floating offshore wind turbines. J. Ocean Eng. Mar. Energy **6**(1), 71–90 (2020). https://doi.org/10.1007/s40722-019-00159-2
19. Zhiping, Z., Chen, J., Liang, H., Zhao, Y., Shao, Y.: Hydrodynamic responses of a 6 mw spar-type floating offshore wind turbine in regular waves and uniform current. Fluids **5**(4), 187 (2020)

Developing a Decision Support System for a Pumped Storage Hybrid Power Plant

Ilkay Ekici(✉), Feyza Durgut, and Gökçen Uysal

Department of Civil Engineering, Eskişehir Technical University, 26555 Eskişehir, Türkiye
{ilkayekici,feyzadurgut,gokcenuysal}@eskisehir.edu.tr

Abstract. Investments in renewable energy systems have been increasing in recent years in order to reduce climate-dependent fossil fuels and to supply the increased energy demand. One of the main difficulties with renewable resources is that they depend on meteorological conditions and produce intermittent energy. Energy storage systems are being developed to preserve the excess energy produced and to use it in situations where energy production is less than expected. Pumped Storage Hydropower Plants (PSHPs) have an important role to overcome these challenges. Operating PSHP in a hybrid manner together with another renewable energy source (solar, wind, etc.) becomes more advantageous for improving system efficiency. In this study, a hybrid system is developed as a Decision Support System (DSS) that wind and solar energy separately are the main energy generator, which stores the excess energy produced with PSHP. The energy production of the Wind Power Plant (WPP) is calculated using wind speed data acquired by the ERA5 global dataset. The energy production of the Solar Power Plant (SPP) is calculated by the solar radiation and cloudiness data. The study also investigates the effects of the developed hybrid system on both observed and predicted market prices. Hourly market prices are predicted via the Long Short-Term Memory (LSTM) model using the lagged market price data sets. According to the results, the hybrid system is superior to the conventional WPP/SPP regardless of observed and predicted market prices. The study is a pioneer application in Türkiye and provides a DSS to use energy resources management more efficiently by increasing the operational flexibility for the decision-makers. The preliminary results also promising as the total revenues increased by around %5.7 and %12.5 for the Hydro-Wind Hybrid System (HWHS) and the Hydro-Solar Hybrid System (HSHS), respectively.

Keywords: Pumped-storage hydropower plant · Wind energy · Solar energy · Machine learning · Market price prediction

1 Introduction

Energy demand is increasing continuously and yet energy production in many countries is highly dependent on limited non-renewable resources. Their consumption is also the major reason for greenhouse gas emissions. The key solution is to increase the installed power of renewable energy sources such as hydro, wind, and solar. Their operational

S. Haddout et al. (Eds.): CCORE 2022, SPEES, pp. 43–63, 2023.
https://doi.org/10.1007/978-3-031-26967-7_4

flexibility in the energy markets is rather restricted because they provide intermittent energy generation and are of highly dependent on meteorological conditions [1].

For example, solar energy production depends on the location of the panels, the cloudiness, and the season of the current year. Also, wind energy is mainly based on the available wind speed.

On the other hand, Pumped Storage Hydroelectric Power Plant (PSHP) [2] is one of the large-scale and effective energy storage alternatives in the form of potential energy. PSHP facilities stand out from other storage systems with their high capacity, long-economic life, sustainability, and high efficiency [3]. According to IHA's Hydropower Status Report [4], the total installed PSHP capacity for the entire world (in 2019) was estimated at 158 GW, and it was expected to increase by about 50 percent to about 240 GW by 2030.

The main operating principle of PSHP is to store water in the upper reservoir (using pumps) when electricity demand is low and to produce energy via turbines from stored water when electricity demand is high. A typical PSHP recovers 70–80% of the input energy. The main reason for 20–30% energy loss is based on friction losses in waterways and pump-turbine equipment [5]. Despite the losses, the profit is gained because there is a chance to produce energy at peak hours in such systems [6].

The hybrid PSHP system design is based on using energy for pumping produced from variable and uncontrolled other renewables (wind and solar) (in other words storing the energy). So that their (wind and solar) entry into the energy markets becomes easier. Depending on the main producer in hybrid systems they are called either Hybrid Wind-Hydro System (HWHS) or Hybrid Solar-Hydro System (HSHS).

There is a number of examples of the implementation of hybrid PSHP systems in the literature. Each of these studies considers different aspects of the system issues such as feasibility, economics, reliability, techno-economic, energy management, etc. [7]. For example; Tuohy and Malley (2011) [8] investigate the operation of the Irish power system with wind energy, considering with and without PSHP. Ma et al. (2014) [9] perform an hourly simulation of a hybrid solar-wind and PSHP system for a remote island in Hong Kong. They demonstrate that the interruption of renewable resources can be compensated by PSHP. Yıldız and Şekkeli (2016) [10] develop an optimization algorithm using linear programming to maximize the day-ahead market bids. Instead of the stand-alone WPP, the annual income is increased by the hybrid PSHP system. Yıldıran and Kayahan (2018) [11] point out the insufficient number of studies related with the real-time operation of hybrid systems and propose a risk-averse stochastic model predictive control technique. Rathore and Patidar (2019) [12] assess the reliability of hybrid PSHP using probabilistic modeling. Ercan and Kentel (2022) [13] develop an optimization model to maximize daily revenue for a HWHS by implementing 25 MW installed capacity of the WPP.

Accurately estimation of the (non-stationary and nonlinear) electricity prices in the spot market is one of the main issues in the PSHP operation. Traditional Artificial Neural Networks (ANNs) inspired is used for decades to solve many time-series predictions. Recurrent neural networks (RNN) (a type of ANN) use their memory to process variable-length input strings and exhibit temporal dynamic behavior. Thus, RNNs are capable of executing more complex calculations [14]. During standard RNN training, some problems such as the disappearing or bursting of gradients may occur [15]. These problems reduce the RNN's ability to learn long-time relationships in the input models.

To avoid these problems, Hochreiter and Schmidhuber (1997) [16]develop a special type of RNN called Long-Short Term Memory (LSTM. LSTM networks are a modified version of RNNs that make remembering historical data in memory easier. For example; Feng et al. (2019) [17] apply LSTM for the controller optimization problem of the PSHP. They demonstrate the supremacy the LSTM model against autoregressive model with exogenous variables, support vector machine, and feedforward neural network. Another study [18] offers a hybrid model which combines adaptive wavelet transform (AWT), LSTM and random forest (RF) algorithm to predict the electricity production in hydroelectric power plant. Ercan and Kentel (2022) [13] apply the LSTM network to predict electricity prices in the day-ahead spot electricity market.

Although Türkiye has a potential of implementing PSHP, however there is no PSHP in operation in Türkiye. Barbaros (2019) [19] indicates the potential of Gökçekaya dam for serving the peak demand and load balancing. In this study, Gökçekaya Dam is selected as the application area considering its potential [20]. Rather than a conventional setup, the PSHP is designed in a hybrid manner under two different renewable (wind and solar) integration. Both hybrid systems are also analyzed using observed and the LSTM model-based predicted market prices to simulate the effects of the systems in an hourly real-time operation.

2 Methodology

2.1 Electricity Production Using Hybrid Systems

Hydro-Wind Hybrid System (HWHS) is defined as the hybrid operation of a Wind Power Plant (WPP) and PSHPs in a joint operation. Hydro-Solar Hybrid System (HSHS) is defined as the hybrid operation of a Solar Power Plant (SPP) and PSHPs in a joint operation. A typical schematic for HWHS and HSHS schematic is shown in Fig. 1. In a hybrid system, the energy produced from other renewable energy sources is used for pumping and upper reservoir water storage is used to generate energy at specified hours when mostly energy requirement increases.

In this study, the HWHS and the HSHS are computed separately, and the simulations are conducted as shown in Fig. 2. For both cases, the meteorological data sets are generated, the corresponding energy either from the WPPs or the SPPs are calculated, and the generated energy from wind/solar renewables is directly used for the pumping from the lower reservoir (Gökçekaya dam lake) to the upper reservoir, which is also designed. The water stored in the upper reservoir is turbined when the energy demand so that the energy prices are increased. Moreover, the market prices are predicted within the LSTM framework for both cases and the system revenues are compared using the observed and predicted prices. The revenue of the hybrid systems is also compared with the conventional renewable (wind/solar) operation.

WPP should be operated when the wind speed is available enough during the course of the day. Similarly, SPPs cannot be continuously operated. The water extracted from the lower reservoir lake is pumped into the upper reservoir during the production of

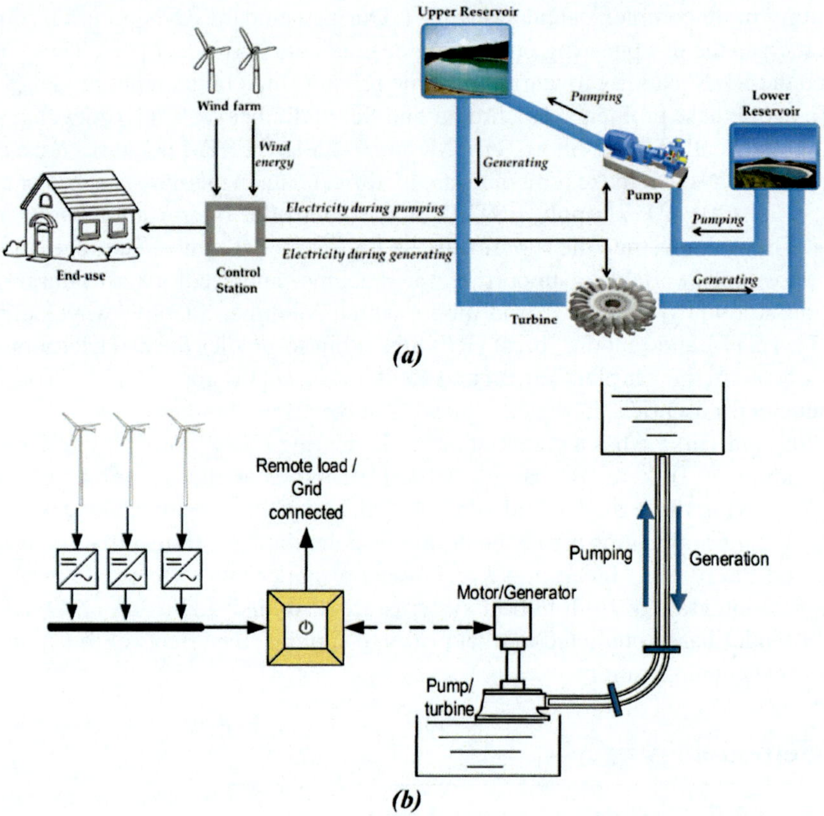

Fig. 1. Schematic of a typical hybrid system (a) Hydro-wind hybrid system [21], (b) Hydro-solar hybrid system [7]

electricity from the renewables that result in filling the volume of the upper reservoir. In the study, the lower reservoir water level is assumed to be constant since it has a higher capacity compared to the upper reservoir. As the volume of the upper reservoir increases, the head of the system also increases. Also, the pump discharge hourly changes depending on the energy coming to the system and the head.

The simulations of the hybrid systems are accomplished with the mass balance and pump/turbine equations (Eqns. 1 – 4). The upper reservoir is designed as an artificial pool, thus the relationship between the upper reservoir water level and its volume is assumed to be linear.

$$Q_i^p = \frac{E_i^{w/s}}{H_i \times e \times \gamma \times \Delta t} \tag{1}$$

$$V_i^p = V_{(i-1)}^p + Q_i^p \times \Delta t \tag{2}$$

$$V_i^g = V_{(i-1)}^g - Q_i^t \times \Delta t \tag{3}$$

$$E_i^g = Q_i^t \times H_i \times e \times \gamma \times \Delta t \tag{4}$$

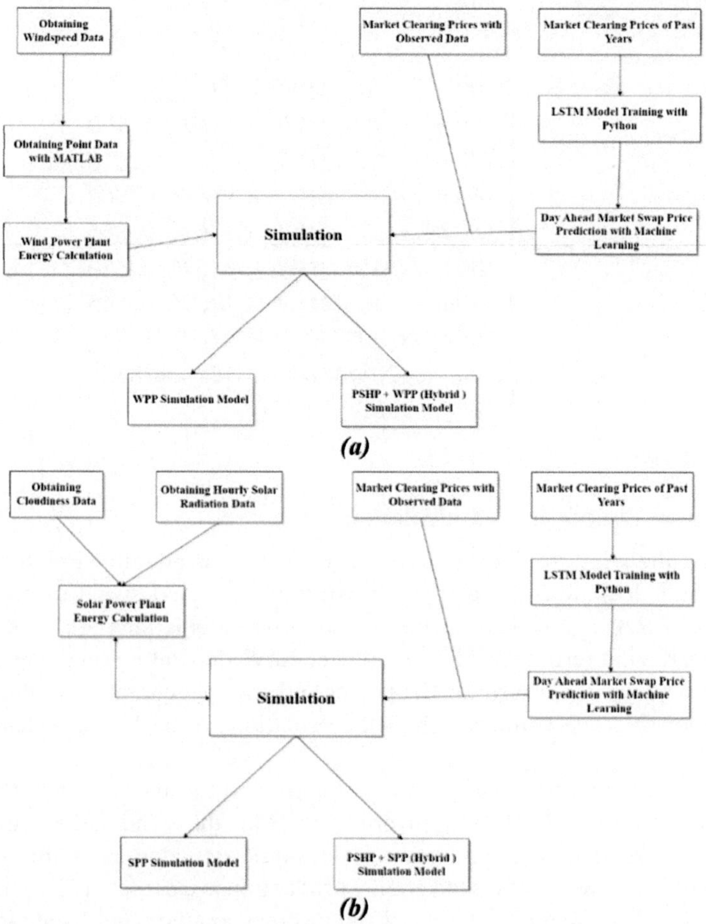

Fig. 2. The flowchart of the study (a) Hydro-wind hybrid system (HWHS) configuration (b) Hydro-solar hybrid system (HSHS) configuration

where Q_i^p (m^3/s), is the pump discharge; $E_i^{w/s}$ (Wh), is the energy production by WPPs/SPPs; H_i (m), is the gross head between the lower (existing dam reservoir lake) and the upper reservoir levels (designed within the context of the study); e, , is the total (hydraulic, turbine, generator,etc.) system losses; γ (N/m^3), is the specific weight of the water; Δt (hour) is the simulation time step; V_i^p (m^3), is the upper reservoir volume during pumping; V_i^g (m^3), is the upper reservoir volume during energy generation; Q_i^t (m^3), is the turbine discharge; E_i^g (Wh), is the energy generation from the upper reservoir; i is the time index (hour).

The turbine discharge is subject to operational decisions. Since energy generation is more profitable when Market Clearing Price (MCP) values are higher, in this study the energy generation decision (so that the turbine flow values) is based on the empiric equation (Eq. 5).

$$Q_i^t = \begin{cases} 190m^3/s, 90\$/MWh < MCP_i \\ 180m^3/s, 70\$/MWh < MCP_i < \$90 \\ 170m^3/s, 60\$/MWh < MCP_i < \$70 \\ 160m^3/s, 55\$/MWh < MCP_i < \$60 \\ 150m^3/s, 50\$/MWh < MCP_i < \$55 \\ 140m^3/s, 45\$/MWh < MCP_i < \$50 \\ 130m^3/s, 40\$/MWh < MCP_i < \$45 \\ 100m^3/s, 30\$/MWh < MCP_i < \$40 \\ 50m^3/s, 20\$/MWh < MCP_i < \$30 \\ 20m^3/s, MCP_i < \$20 \end{cases} \tag{5}$$

2.1.1 Hourly Wind Energy Calculations

Considering the wind turbine can adjust itself for the available wind direction, hourly wind energy calculations are computed using hourly wind speed data obtained from global data (ERA5) sets. The details of the wind calculations are described in Data Section. Each wind turbine has its own power curve depending on the technical details of the turbine. One can determine the power and energy generation of the wind turbine using the power curve without technical calculations of each component of the wind turbine.

In this study, the power curve of a selected wind turbine, a General Electric (GE) 2.5 MW wind turbine [22], is used provided from the wind turbine manufacturers' catalog. The corresponding wind power values are calculated using the velocity power relationship. Then, the hourly wind energy production is calculated (Eq. 6). In this study, the time interval is equal to one hour. Wind turbines, on the other hand, are designed to be operated for 11 h every day between 08:00 and 18:00.

$$E_i^w = P_w \times \Delta t \tag{6}$$

where $\boldsymbol{E_i^w}$ (Wh), is the produced energy by WPP; $\boldsymbol{P_w}$ (W), is the available wind power; $\boldsymbol{\Delta t}$ (hour) is the simulation time step; $\boldsymbol{i}$ is the time index (hour).

2.1.2 Hourly Solar Energy Calculations

Solar energy panels can have different installations such as land SPP, floating SPP, etc. Land SPP is installed on the ground with steel constructions on large lands. There are also floating SPPs, which are positioned on the water body surfaces and prevent the water from evaporating. At the same time, the water cools the panels, and the life of the panels is extended, thus the panels work more efficiently. In this study, solar panels are located in three different regions.

The two regions are located above the upper and lower reservoirs (floating type) and the others (land type) are assumed to be installed at the nearby region. The energy potential that can be produced from solar panels for storage in the PSHP is calculated separately for each area as expressed in Eq. 7.

$$E_i^s = R \times \mathrm{e} \times \mathrm{A} \times \frac{100 - C}{100} \tag{7}$$

where E_i^s (Wh), is the produced energy by SPP; R, , is the incoming radiation (Wh/m^2); e, is the efficiency of solar panels; A, , is the area of total solar panels (m^2); C, , is the cloudiness (%).

In this study, the efficiency of the different solar panels (depending on land or floating type) is also separately considered.

2.2 Electricity Spot Market Price Forecasting with LSTM

Hourly MCPs are estimated using a machine learning LSTM model. An ordinary LSTM unit contains four different layers that communicate with each other. These layers are; a cell, an input gate, an output gate and a forget gate. The cell remembers values in variable-length time intervals, and these three gates regulate the flow of information into and out of the cell. An example of this typical LSTM architecture schematic is given in Fig. 3.

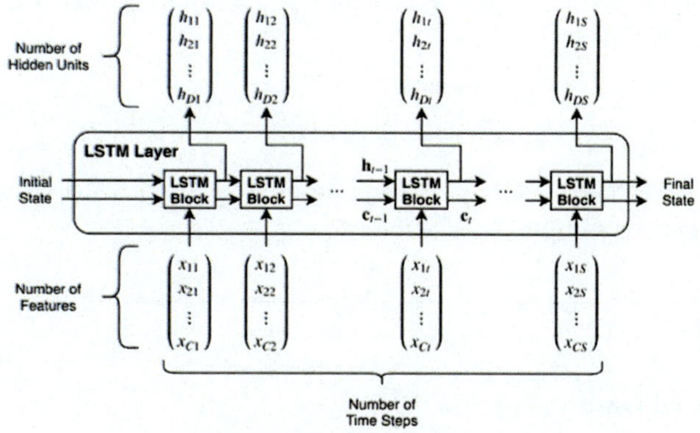

Fig. 3. LSTM diagram [23]

The developed LSTM model employs the lagged MCPs of the previous time series data set and the model predicts one hour beforehand. An equation summarizing the input and output relationship is given in Eq. 8.

$$y_{t+1} = f(y_{t-1}, y_{t-2}, y_{t-3}, \ldots, y_{t-24}) \tag{8}$$

Before using LSTM model, the model is trained to find the optimum hyperparameters. These parameters are set after a trial-and-error procedure.

In our model, the mini-batch size which defines the number of subsamples is carefully chosen as 128. The model is retrained depending on the epoch number with the new training set and the weights are updated again. This process is repeated at each training step to try to calculate the most appropriate weights for the model. The best model performance is achieved with 100 epoch number.

The number of hidden units corresponds to the amount of information remembered between time steps (hidden state). The hidden state can contain information from all previous time steps, regardless of the sequence length. The number of hidden units selected is 4 and "ADAM Optimizer" is chosen as the optimizer algorithm. ADAM can be used instead of the classical stochastic gradient descent procedure to iteratively update network weights based on training data.

Momentum helps to detect the direction of the next step with the knowledge of the previous steps. In our model, the momentum value is set as 0.900. The learning rate is also another hyperparameter that controls how much the model is modified in response to the estimated error each time the model weights are updated. The initial learning rate selected is 0.050, and the learning rate decreases by 0.100 every 10 epochs.

There are 446 different packages for training and the weights are updated after each package. As a validation criterion, the number of early stops is chosen as 25. The input data is subjected to the normalization process between 0 and 1 (Eq. 9).

$$z = \frac{x_i - \mu}{\sigma} \quad (9)$$

where; x_i, is the actual value, μ, is the mean, and σ is the standard deviation.

"Mean Squared Error (MSE)" is selected as the loss function (Eq. 10).

$$MSE = \frac{1}{n}\sum_{i=1}^{n}(xi - yi)^2 \quad (10)$$

where, x_i is the observed MCP, y_i is the output from the LSTM network, and n is the number of outputs or samples in a batch.

3 Data Sets

3.1 Wind Speed Data

Wind speed varies depending on location and time. In this study, wind speed data (2019–2020 water year) are obtained from the ERA5 dataset of the European Center for Medium-Term Weather Forecasts (ECMWF) for the selected study area. The ERA5 dataset is used for similar studies in the literature [24,25, 26].

The spatial resolution of the data is 0.25° x 0.25°. ERA5 wind speed data is provided in netCDF4 file format. These data sets contain five parameters: time, latitude, longitude, eastward wind speed at 10 m (U10) and northward wind speed at 10 m (V10). The data is extracted and processed by a script using MATLAB. The resultant wind speed data are calculated using the Pythagorean Theorem.

Wind speeds obtained from the ERA5 dataset are 10 m above the surface. Therefore, they require an extrapolation concerning the hub height of the wind turbine. The hub height is also based on the manufacturer. In the study, a General Electric GE 2.5 - 120 wind turbine with a hub height of 139 m is preferred. The power law equation (Eq. 11) is used to extrapolate the wind speed to the hub height. The power law provides to calculate the vertical wind speed profile [27].

$$\frac{U(z)}{U(zr)} = \left(\frac{z}{zr}\right)^{\alpha} \tag{11}$$

where $U(z)$ (m/s), is the wind speed at altitude z (m); $U(z_r)$ (m/s), is the reference wind speed at reference altitude (m), and α is the force law exponent.

Power law exponent varies with many parameters such as altitude, time, season, location, wind speed, temperature and various thermal and mechanical parameters. However, these parameters complicate the power law exponent, thus reducing the simplicity and applicability of the power law[27]. Therefore, an empirical equation proposed by Justus (1978) [28] is used in this study (Eq. 12). The total of 200 turbines are used for wind energy production.

$$\alpha = \frac{0,37 - 0,088 * ln(U(zr))}{1 - 0,088 * ln\left(\frac{zr}{10}\right)} \tag{12}$$

3.2 Solar Radiation and Cloudiness Data

In this study, in order to obtain the solar energy to be used in the simulation model, the spatial solar radiation data is calculated with the ArcGIS program using solar radiation analysis tools. To that end, the SRTM Digital Elevation Model (DEM) is used provided by USGS's EarthExplorer platform [29]. The solar panel regions are selected on the DEM and designed. 120,000 solar panels with 21.6% efficiency for ground SPP, 20,000 floating solar panels with 17.09% efficiency for the lower reservoir and 200,000 floating solar panels with 17.09% efficiency for the upper reservoir are used in the selected areas.

Another important variable that negatively affects the energy production performance of solar panels is the cloudiness/cloud cover ratio. Solar radiation measurements do not provide information about the shadow factor [30]. For this reason, the cloud cover ratio in the region where the panels will be located must also be known for the calculation of the energy to be produced. In this study, cloudiness data is obtained from a nearby meteorological station.

3.3 Market Clearing Price (MCP) Data

In the energy sector, all stages from electricity generation to distribution are supervised and managed by various institutions. Electricity generation activities by the Electricity Generation Company (EÜAŞ), electricity transmission activities by the Turkish Electricity Transmission Company (TEİAŞ), and then electricity distribution by private electricity distribution companies are carried out.

In order to carry out the generation, transmission, distribution, purchase, and sale of energy in the right amount at the right time and to create competition, a market operated by the Electricity Markets Operation Joint Stock Company (EPİAŞ) has been created. There is a synchronization between energy production and consumption, and it is kept in balance by EPİAŞ.

Complete substitution of electrical energy is not possible, and it is costly to store, so the electricity market has different characteristics from other product markets. Electricity supply and demand, which varies according to the amount of use in the industrial sector, causes fluctuations in electricity prices and costs [31]. Irregularities in the energy market are an undesirable situation as it prevents access to quality, reliable and continuous energy. To achieve this balance, the amount of energy to be given to the grid by the producers is kept under control by placing the Market Clearing Price (MCP) and System Marginal Price (SMP), which are determined in line with the supply and demand. Hourly MCP and SMP data are provided from the EPİAŞ Reporting Platform.

Hourly electricity price data are converted to US dollars with exchange rates obtained from the Electronic Data Distribution System (EVDS) platform of the Central Bank of the Republic of Turkey [32]. The data period is between July 2009 to December 2021.

Data from July 2009 to January 2016 is used to train the LSTM network. These data are divided into three different categories. These are the training period, the validation period, and the testing period. Training data includes data from July 2009 to December 2015. Validation data includes data from January 2016 to December 2019. The test data includes the data between January 2020 and December 2020. Figure 4 depicts the monthly average of the hourly data set in terms of MWh.

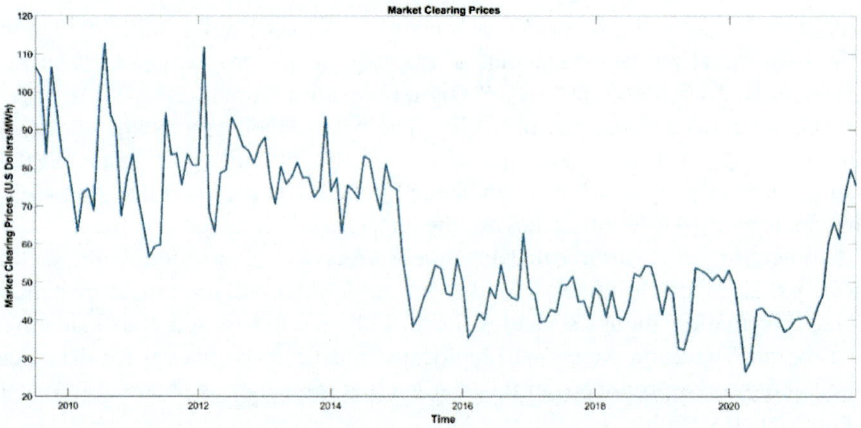

Fig. 4. Average monthly MCP data set

4 Study Area

In this study, a hypothetical HWHS and a hypothetical HSHS are designed to demonstrate the application of the developed models. Although there is no PSHS in operation in Türkiye yet, the Gökçekaya Dam (located at Sakarya River Basin) has been selected as the study area since it is one of the best locations considering turbine design flow and available head in a feasibility study [20].

Gökçekaya Dam does not have an active upper reservoir; thus, the location and the dimensions of the upper reservoir are designed within the content of this study. While designing the upper reservoir, we use the storage capacity equation for PSHPs in Eq. 13 [33]. The initial volume of the reservoir is assumed to be the dead volume (5% of the total volume).

$$S = \frac{976 \times C \times t_s}{H \times e_g} \tag{13}$$

where S (acree-feet), is the storage capacity of the upper reservoir; C (MW) is the maximum power plant capacity; t_s (hours) is the operation duration (hour); H (feet) is the gross head; e_g is the efficiency (assumed as 0.85).

In the study, the DEM data is also used to select the appropriate location for the upper reservoir with several scenarios, and the remaining calculations are given for one scenario in the paper. Since wind speeds increase with high elevation, the location of the WPPs is selected for the nearby region. SPPs are installed mostly on southern sloping slopes in the Northern Hemisphere, thus land-based solar panels are assumed to be installed within that concept. Also, extra floating solar panels are installed on the reservoir lake to save from land use.

5 Results

5.1 Data Analyses

5.1.1 Wind Data Analyses

The hourly ERA5 wind speed data extrapolated to hub height is given for the 2020 water year in Fig. 5. According to the graph, the wind speed starts to increase around 08 AM and reaches its peak average value around 01 PM and it gradually decreases through the day. The minimum mean value is observed at 5 AM.

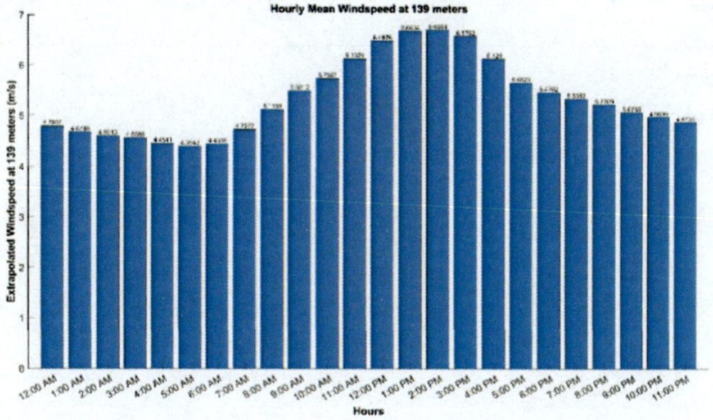

Fig. 5. Extrapolated windspeed data

Figure 6 shows the variation of the hourly wind speed data in the boxplot-wishers diagram. The diagram shows the different percentages of the data set, so its distribution for each hour can be evaluated. The median values are higher between 12 PM – 03 PM. Outliers are expected from 08 AM – 10 AM and 07 PM – 11 PM.

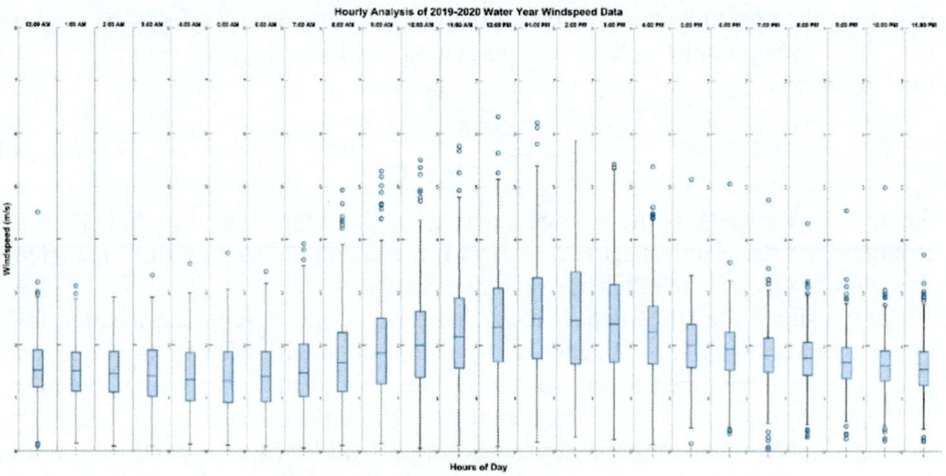

Fig. 6. Windspeed data analysis

5.1.2 Solar Data Analyses

In Fig. 7, the calculated solar radiation data for three different SPP regions are given as an annual time series. In general, high radiation is observed in the summer months and this value decreased towards the winter months.

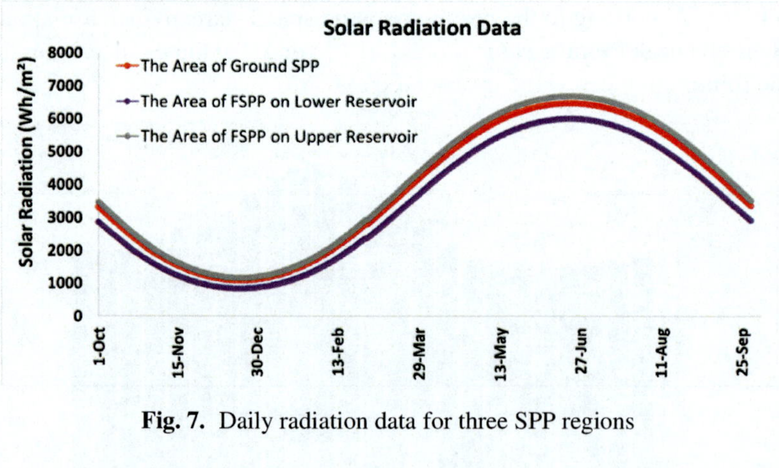

Fig. 7. Daily radiation data for three SPP regions

5.1.3 MCP Data Analyses

Figure 8 demonstrates the hourly average values of the observed MCP for July 2009 – December 2021 time period. Usually, the MCP values are low between 03 AM and 06 AM. The highest value is obtained at 11 AM. In the afternoon, the prices are generally higher than the night prices.

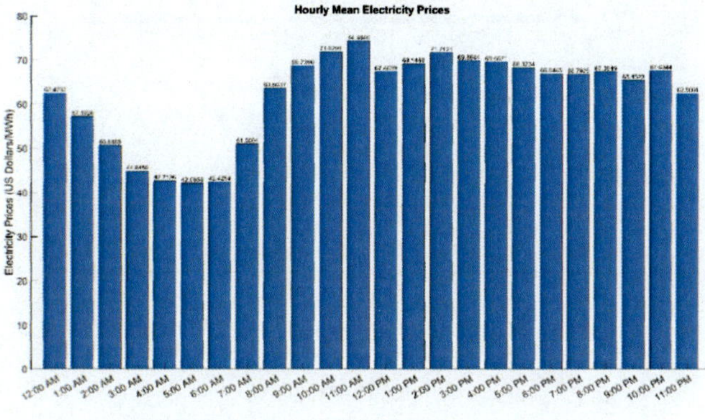

Fig. 8. Hourly mean electricity prices

6 LSTM Results

In this part, the predicted LSTM model results are compared with observed electricity MCP data.

The comparison of the observed and predicted MCP data for the 2020 water year is given in Fig. 9. Mostly the predicted MCP values are close to the observed values.

The performance of the LSTM can be low for outliers. MCP values are also depicted by boxplot and whisker charts for each hour within the course of the day in Fig. 10. In the graphs, the red boxplots represent the hourly predicted MCP data, while the blue boxplots represent the observed MCP data.

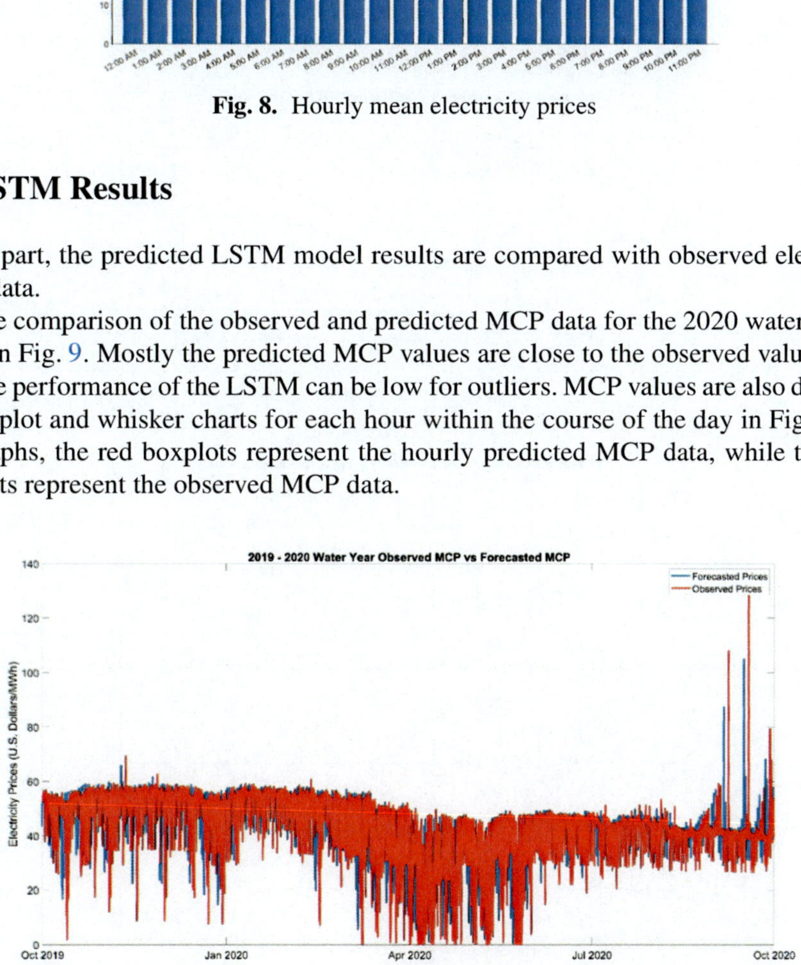

Fig. 9. Observed and predicted MCP for the 2020 water year

According to Fig. 10, the estimated MCP for 12–02 AM has a lower median value, while outliers are higher. The medians of both observations and predictions are close to each other between 03–04 AM, but the predicted MCP has a larger spread. The higher median values are detected between 05 – 07 AM, but the observed MCP shows longer whiskers. The observed and predicted values and medians between 08–10 AM are close to each other.

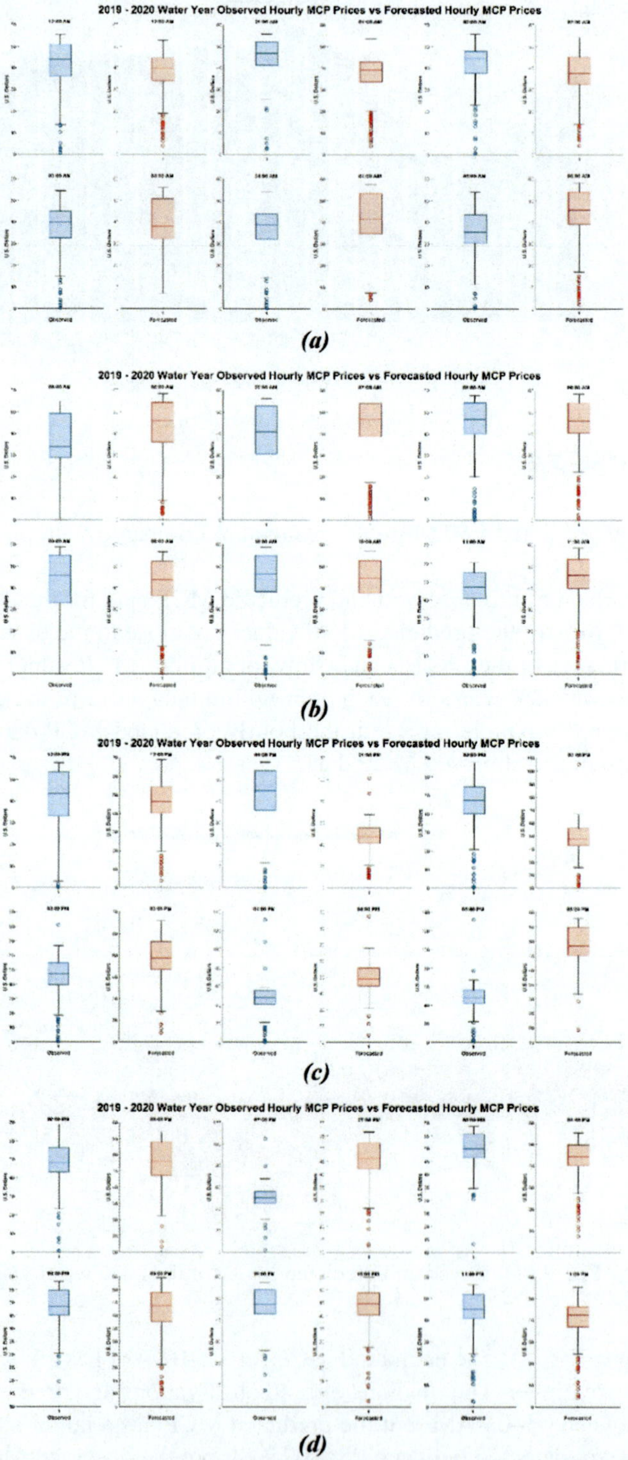

Fig. 10. Hourly observed and predicted MCP boxplot and whiskers diagram comparison for 2020 water year (a) 12 AM–05 AM, (b) 06 AM–11 AM, (c) 12 AM – 05 PM, (d) 06 PM–11 PM

On the other hand, the medians and correlations are highest at 11 AM, but the predicted MCP has higher outliers. The median of the predicted values at 12 PM is slightly higher than the median of the observed values, with the outliers of the predicted values reaching much higher values.

Between 01–02 PM the medians are close, but the distribution is wider for the observed MCP, with a very high outlier at 02 PM. At 03 PM, the medians are close, the spread is large for the predicted MCP boxplot, and there is a high outlier in the observed MCP.

Although the whiskers of the box plots are different between 04–05 PM, the medians are close to each other. The data distributions and medians are close to each other between 06–07 PM, but the lower and the upper outliers are high for predictions and observations, respectively.

The median and spread show similar behaviour at 08 PM, but more outliers are detected in the predicted MCP. The median of the predicted MCP is slightly lower than the observation median at 09 and 11 PM. The distribution and median values have a higher correlation at 10 PM.

In general, overall the predicted values are quite consistent, although different outliers and some deviations are observed.

6.1 Hybrid Systems

In this part, the electricity generations from both hybrid systems are discussed and compared with the single WPP/SPP systems. The energy produced from the wind/solar is stored in the upper reservoir by the pumps and the storage is used later for high demand/MCP hourly.

Therefore, two energy production scenarios are created for both hybrid systems using observed and predicted MCP. The results are demonstrated for the 2020 water year.

6.1.1 Hydro-wind Hybrid System (HWHS) Results

The total hourly energy production by the WPPs is presented for the 2020 water year in Fig. 11. The WPPs are operated for 11 h in a course of a day. The inter-seasonal variability of the energy production from the WPP is due to the available wind speed in actual application.

Since the energy generation from the upper reservoir decision is based on the demand so the MCP data set (Eq. 5), two different energy production is calculated considering the observed and predicted MCP as shown in Fig. 12. Annual total revenues (selling to the grid) from HWHS are calculated as \$16.84 million and \$19.67 million, for observed MCP and predicted MCP.

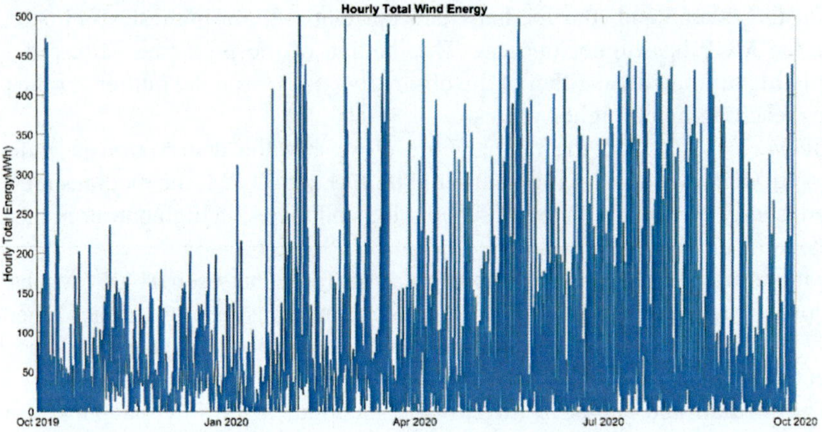

Fig. 11. Wind energy production using WPPs for the 2020 water year

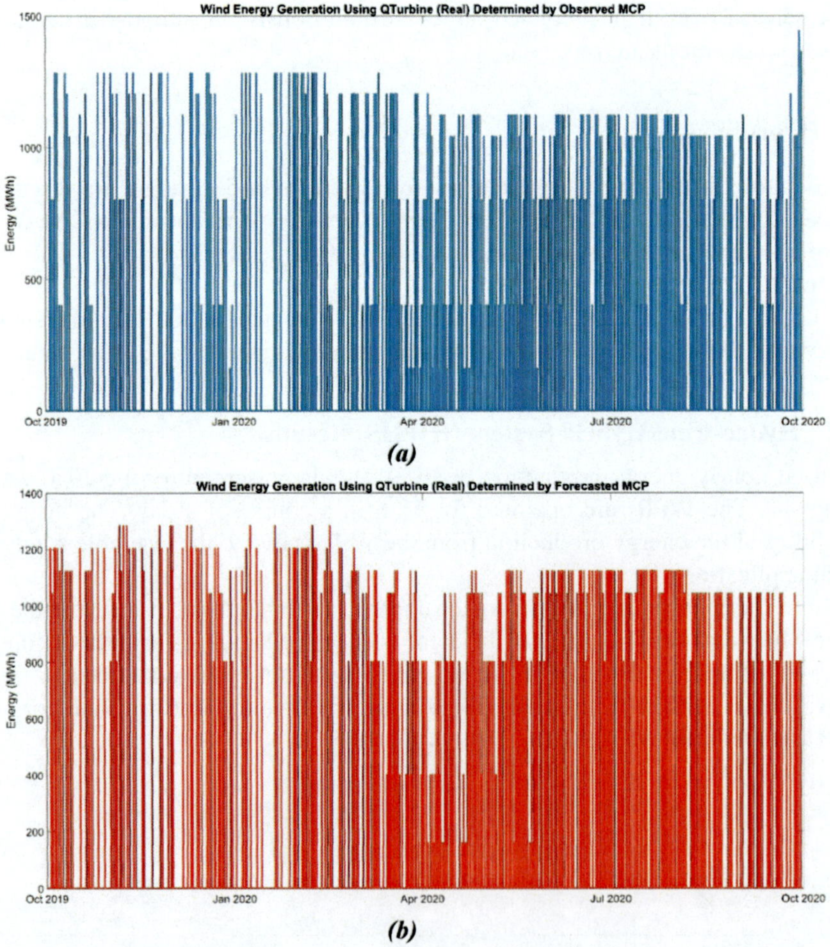

Fig. 12. Energy generation from HWHS (a) Observed MCP (b) Predicted MCP

6.1.2 Hydro-solar Hybrid System (HSHS) Results

The energy production potential of solar panels is calculated using the solar radiation data sets. The SPP energy production is demonstrated in Fig. 13. The SPPs are operated for 8 h over a course of a day. The variability of the energy production from the SPP is quite different in each season of the time due to mainly cloud and received solar radiation. Since the energy generation from the upper reservoir decision is based on the demand so the MCP data set (Eq. 5), two different energy production is calculated considering the observed and predicted MCP as shown in Fig. 14. Annual total revenues (selling to the grid) from HSHS are calculated as \$73.32 million and \$72.77 million, for observed MCP and predicted MCP.

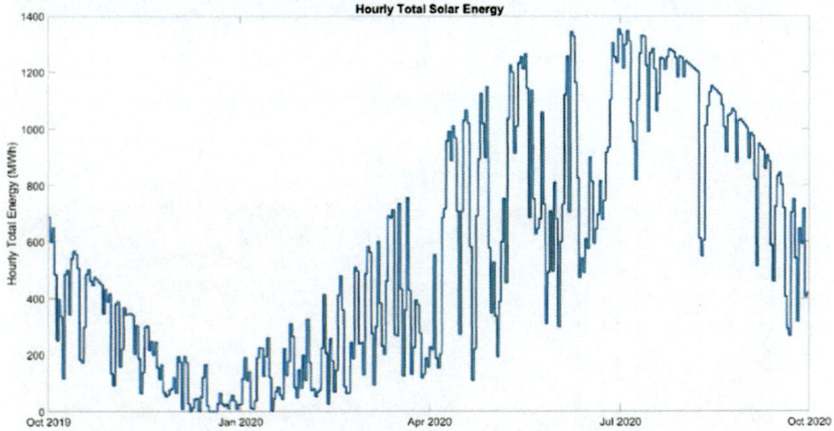

Fig. 13. Solar energy production using SPPs for the 2020 water year

6.1.3 Comparison of the Results

In order to make a fair comparison, a benchmark model for both cases is generated as only WPP or only SPP without integration into a PSHP. The summary of the all results is given in Table 1. According to the comparison, hybrid operations provide to generate more revenue and system stability since the generation is not directly based on meteorological/environmental constraints.

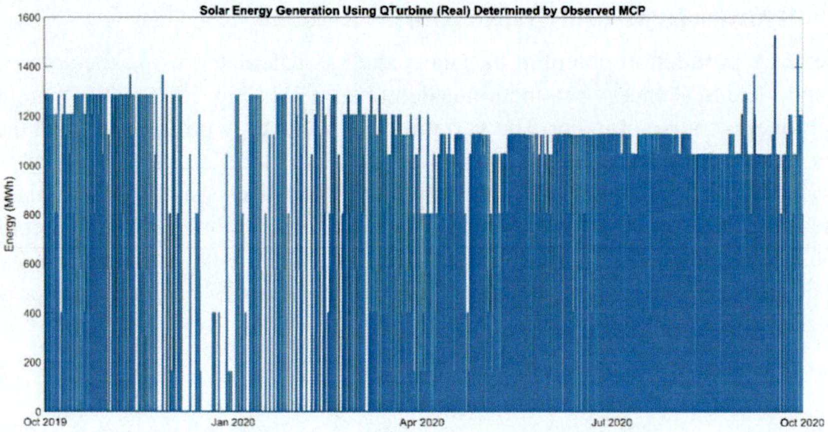

(a)

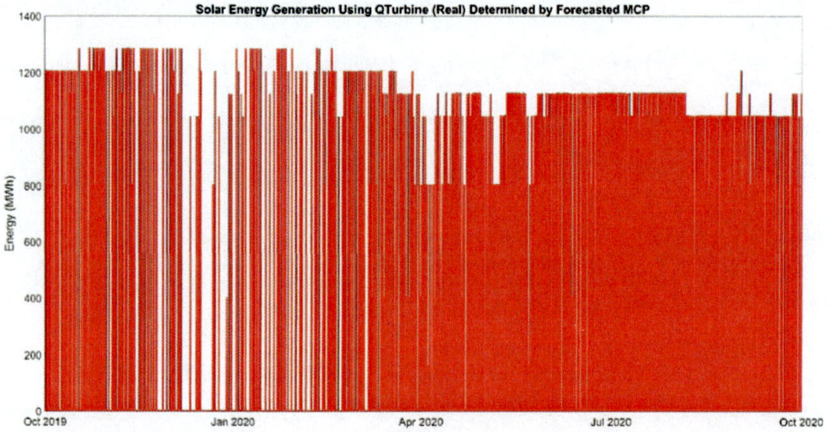

(b)

Fig. 14. Energy generation from HSHS (a) Observed MCP (b) Predicted MCP

Table 1. Comparison of the results

System	MCP	Revenue (million $)
HWHS	Observed	16.90
	Predicted	18.25
Conventional WPP (Benchmark)	Observed	15.94
	Predicted	16.83
HSHS	Observed	73.32
	Predicted	72.77
Conventional SPP (Benchmark)	Observed	64.13
	Predicted	69.65

7 The Limitation of the Study

The basic assumption of this study is to implement a hypothetical hybrid PSHP on Gökçekaya Dam. The upper reservoir (a new artificial pond) is designed according to a selected operation time duration, available head, and maximum turbine flow provided by the initial feasibility studies. The locations of SPPs (land or floating type), the number of SPPs, the number of WPPs, the type of WPPs/SPPs are selected by the authors.

Since there are no designed components (penstock, tailrace, intake tunnel, etc.) of the PSHP in operation, hydraulic losses (friction and minor losses) are assumed and applied in the net head calculations. Therefore, the efficiencies (hydraulic, pump, turbine, transformer, etc.) might differ in other studies.

Even though cloudiness has an areal extent, it is still challenging to represent it by sparse in-situ observations. Due to the limited station network, this study considers one nearby station-based cloudiness data. In the study, also the price estimation LSTM model is still preliminary since it can foresee one hour-ahead market prices. This should be improved up to a day ahead manner.

8 Conclusions

The key solutions to decrease carbon emissions can be satisfied using more efficient systems. However, wind and solar are highly dependent to the meteorological variables; and do not provide operational flexibility. The energy storage systems are essential for a reliable grid stability.

This paper aims to develop and assess the performances of the hybrid systems that are composed of PSHP and other renewables such as WPP and SPP. Firstly, a hypothetical hybrid system is designed for an existing dam reservoir located in Türkiye and operated in integration with WPP and SPP, separately. Secondly, electricity MCP, wind speed data and solar radiation data for the selected area are obtained and analyzed. Then, the upper reservoir is designed and the energy production of the WPP/SPP are estimated. Afterwards, the hybrid systems are simulated under two scenarios. The hybrid operation of the WPP/SPP with the PSHP, and single operation of WPP/SPP. The revenues of both scenarios are obtained with the electricity MPC data observed and predictions using an LSTM network.

The conclusions are summarized as:

- The potential of the SPP for the selected study area is higher than the potential of WPP according to the energy and revenue results.
- By the means of the hybrid system, the produced energy from both renewables is stored in the upper reservoirs and generated for consumption when the electricity MCP is high. This also provides system stability and operational flexibility because meteorological constraints are reduced.
- For both observed and predicted electricity MCP, more revenue is estimated in a hybrid system compared to a single WPP or SPP. The predicted MCP yielded more profits than the observed MCP, in both conditions. This might be attributed to the under- or overestimation of the MCP for highly variable hours.

Here are some suggestions for future work:

- Additional input data, configuration and layers can be tested to improve the performance of the LSTM network.
- In addition to the ERA-5 dataset, the reliability of the data should be checked with gauge-based in-situ wind speed datasets (validation study).
- The economic analysis is not considered within the study. Thus, it would be useful for further studies to include the cost of installation of WPP and SPP together with upper reservoir construction.

Acknowledgment. This study has been supported by TÜBITAK 2209-A - University Students Research Projects Support Program, and Eskişehir Technical University Undergraduate Student Projects (LÖP) 22LÖP180. We would like to acknowledge the Turkish State Meteorological Service (MGM) for data sharing.

References

1. Wang, C.S., et al.: Distributed energy and microgrids (DEM). Appl. Energy **210**, 685–689 (2018)
2. Miller, R.: Opportunities in Pumped Storage Hydropower: Supporting Attainment of Our Renewable Energy Goals By: Rick Miller, P.E., Senior Vice President, Hydropower Division, HDR|DTA and Maureen Winters, Corporate Consultant, Hydropower Division, HDR|DTA, pp. 1–9 (2009)
3. Linden, D., Reddy, T.B.: Handbook of Batteries, Chapter 30, M. Fetcenko, McGraw-Hill, New York, USA (2002)
4. IHA . The world's water battery: Pumped hydropower storage and the clean energy transition, IHA Working Paper 1, pp. 1–15 (December 2018)
5. Priya, K.L.A., Haddout, S., Casila, J.C.C.: Implications of turbulent shear on clay floc breakup along the Atlantic estuary (Bouregreg) Morocco. Int. J. Sediment Res. **37**(2), 248–257 (2022)
6. Bulu, A.: Hidroelektrik santrallerin tasarım ve hesap esasları. Okan Üniversitesi Yayınları, İstanbul (2011)
7. Javed, M.S., Ma, T., Jurasz, J., Amin, M.Y.: Solar and wind power generation systems with pumped hydro storage: review and future perspectives. Renewable Energy **148**, 176–192 (2020)
8. Tuohy, A., O'Malley, M.: Pumped storage in systems with very high wind penetration. Energy Policy **39**(4), 1965–1974 (2011)
9. Ma, T., Yang, H., Lu, L., Peng, J.: Technical feasibility study on a standalone hybrid solar-wind system with pumped hydro storage for a remote island in Hong Kong. Renewable Energy **69**, 7–15 (2014)
10. Yıldız, C., Şekkeli, M.: Türkiyegünöncesi elektrik piyasasındarüzgarenerjisive pompajde polamalı hidroelektrik santraliçin optimum teklifoluşturulması. Pamukkale Üniversitesi Mühendislik Bilimleri Dergisi, **22**(5), 361–366 (2016)
11. Yıldıran, U., Kayahan, İ: Risk-averse stochastic model predictive control-based real-time operation method for a wind energy generation system supported by a pumped hydro storage unit. Appl. Energy **226**, 631–643 (2018)

12. Rathore, A., Patidar, N.P.: Reliability assessment using probabilistic modelling of pumped storage hydro plant with PV-Wind based standalone microgrid. Int. J. Electr. Power Energy Syst. **106**, 17–32 (2019)
13. Ercan, E., Kentel, E.: Optimum daily operation of a wind-hydro hybrid system. J. Energy Storage **50**, 104540 (2022)
14. Garrido, O. a.: Short Term Load Forecasting with Neural Networks, vol. 53(1), pp. 17– 23 (2012)
15. Pascanu, R., Mikolov, T., Bengio, Y.: Understanding the exploding gradient problem, vol. 2(417), pp. 1 (2012). CoRR, abs/1211.5063
16. Hochreiter, S., Schmidhuber, J.: Long short-term memory. Neural Comput. **9**(8), 1735–1780 (1997)
17. Feng, C., Chang, L., Li, C., Ding, T., Mai, Z.: Controller optimization approach using LSTM-based identification model for pumped-storage units. IEEE Access **7**, 32714–32727 (2019)
18. Zolfaghari, M., Golabi, M.R.: Modeling and predicting the electricity production in hydropower using conjunction of wavelet transform, long short-term memory and random forest models. Renewable Energy **170**, 1367–1381 (2021)
19. Barbaros, E.: Feasibility of pumped-storage hydraulic systems based on hourly variation of electricity prices (Master's thesis, Middle East Technical University) (2019)
20. Çiçek, Ö., Özdemir, M.: Örnek Bir Hidroelektrik Santraliİçin Pompaj Depolamalı Hidroelektrik Santrali Tasarımı. Gazi Mühendislik Bilimleri Dergisi **7**(1), 26–35 (2021)
21. Sarasúa, J.I., Martínez-Lucas, G., Platero, C.A., Sánchez-Fernández, J.Á.: Dual frequency regulation in pumping mode in a wind–hydro isolated system. Energies **11**(11), 2865 (2018)
22. General Electric (n.d.). GE Power & Water, Wind Turbine Series 2 .5mw
23. https://www.mathworks.com/help/deeplearning/ug/long-short-term-memory-networks.html (Accessed 28 July 2022)
24. Ruiz, S.A.G., Barriga, J.E.C., Martínez, J.A.: Wind power assessment in the Caribbean region of Colombia, using ten-minute wind observations and ERA5 data. Renewable Energy **172**, 158–176 (2021)
25. Soares, P. M., Lima, D. C., Nogueira, M.: Global offshore wind energy resources using the new ERA-5 reanalysis. Environ. Res. Lett. **15**(10), 1040a2 (2020)
26. Jurasz, J., Mikulik, J., Dąbek, P.B., Guezgouz, M., Kaźmierczak, B.: Complementarity and 'resource droughts' of solar and wind energy in Poland: an ERA5-based analysis. Energies **14**(4), 1118 (2021)
27. Manwell, J.F., McGowan, J.G., Rogers A. L.: Wind Energy Explained: Theory, Design and Application, 2 edn. Wiley (2009)
28. Justus, C.: Winds and wind system performance, Solar Series, Franklin Institute Press (1978). https://books.google.com.tr/books?id=a-RSAAAAMAAJ
29. https://earthexplorer.usgs.gov/. (Accessed 22 December 2021)
30. Kahveci, B.: Dalgacık Analizi Kullanılarak Optik Fotoğraflardan Bulutluluk Oranı Tayini (Doctoral dissertation, Enerji Enstitüsü) (2018)
31. Demirtaş, I., Tarı, E.N.: Türkiye'deElektrik Piyasasının Rekabete Açılmasıve Rekabetçi Elektrik Piyasasında Elektrik Tüketiminin Belirleyicileri: ARDL SınırTesti. JOEEP: J. Emerging Econ. Policy, **6**(1), 148–164 (2021)
32. https://evds2.tcmb.gov.tr/index.php?/evds/archiveMarket/ . (Accessed 24 December 2021)
33. Antal, B.A.: Pumped Storage Hydropower: A Technical Review in the area of Hydrologic and Hydraulic Engineering (May 2014)

A SWOT Analysis for Offshore Wind Energy Development: Turkey Case

Samet Öztürk(✉)

Department of Environmental Engineering, Bursa Technical University, Bursa, Turkey
samet.ozturk@btu.edu.tr

Abstract. The increasing energy need and climate change crisis force countries to seek alternative energy sources. In the last decade, there is a dramatic increase in the investment of renewable energy systems globally. In the renewable energy systems one of the leading energy systems is wind energy. Although current installations of wind energy mostly compose of onshore wind energy systems, offshore counterparts are raising attention recently. In this study, offshore wind development in European and Asian countries is reviewed and a Strength, Weaknesses, Opportunities and Threats (SWOT) analysis for Turkey in terms of an investment on offshore wind is applied. The analysis showed that there are many opportunities and strengths that overcome weaknesses and threats for Turkey to deploy offshore wind on their seas.

Keywords: Europe · Asia · Turkey · Offshore wind · SWOT

1 Introduction

Energy sources are mostly categorized as fossil fuels (e.g. coal, natural gas and petroleum), nuclear power and renewable energy systems (e.g. hydro, wind, solar, geothermal, biomass, wave and tidal). Among the energy sources, renewable energy systems gain more attention recently. There are two main reasons that renewable energy systems are becoming more popular globally. First, climate change force countries to seek alternatives to the fossil fuels which cause global warming. Second, energy need is increasing due to population rise and technological development. One of the leading renewable energy systems is wind energy which compose of 837 GW installation worldwide as of 2021 [1]. Wind energy produced more than 9% of the electricity in 2021 in the World [1]. Figure 1 shows global historic development of onshore and offshore wind energy. In 2001, there was 24 GW of wind energy installation worldwide, while it became 35-fold in 2021 with 837 GW [1]. In 2021, there was 57 GW of offshore wind deployment while onshore counterparts were 780 GW. Another interesting point is that the share of offshore wind in the total wind deployment is increasing steadily. Between 2001 and 2009 the share of offshore wind was only 1% while it increased to 2% between 2010 and 2014. In 2021 there was 7% share of offshore wind in all wind deployment. This shows that offshore wind deployments increased more than onshore

S. Haddout et al. (Eds.): CCORE 2022, SPEES, pp. 64–71, 2023.
https://doi.org/10.1007/978-3-031-26967-7_5

counterparts recently. In this study, the development of offshore wind is reviewed, especially for leading countries in Europe and Asia. Furthermore, as a case study, Turkey is analyzed by applying Strength, Weaknesses, Opportunities and Threats (SWOT) method in terms of an investment on offshore wind. The research questions for this study are in the following:

- How European and Asian countries achieved high deployment in offshore wind?
- What are the strengths, weaknesses, opportunities and threats for Turkey which has no offshore wind currently to invest on offshore wind?

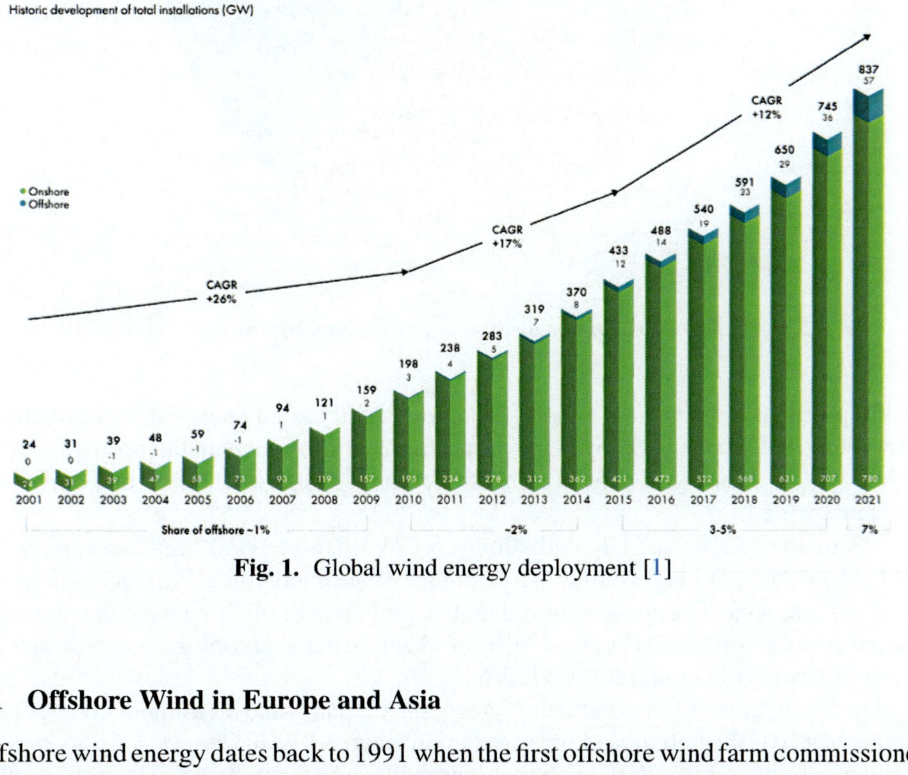

Fig. 1. Global wind energy deployment [1]

1.1 Offshore Wind in Europe and Asia

Offshore wind energy dates back to 1991 when the first offshore wind farm commissioned in Denmark. Until 2020's European Countries were the leaders of offshore wind in terms of capacities. However, starting with 2020 and improving in 2021, China took over the leadership of the offshore wind installation with almost 30 GW total capacity increase in only two years. This was a dramatic pace considering a total cumulative installation of 48 GW by the end of 2021 [1].

There are eighteen countries, of which only the US is out of Europe and Asia, that utilize offshore wind currently. The US has only 42 MW installation capacity, thus in this study the developments of offshore wind in the US is ignored. Figure 2 shows the installation capacity distribution by the countries as of the end of 2021 [1].

In Europe the leading offshore country is United Kingdom with 12 GW installation as of 2021 [1]. UK, which is an island, is surrounded by the seas. There is enormous

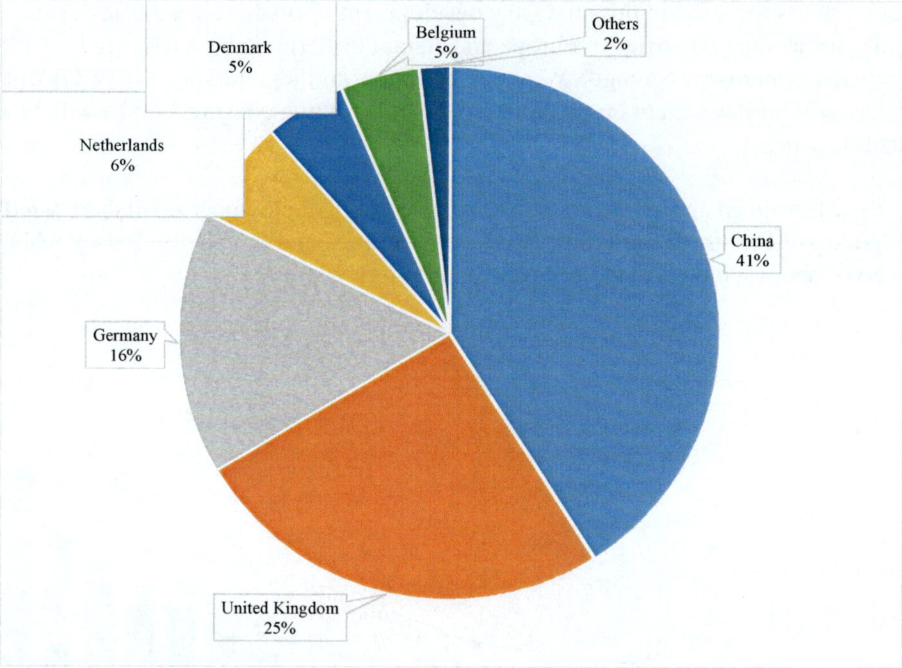

Fig. 2. Cumulative installation distribution by the countries as of the end of 2021 [1]

wind potential on their near shores. Also, the country is an energy dependent country. Therefore, the offshore deployment was pushed by these two main factors in the country. Moreover, high feed-in-tariffs up to 0.246 $/kWh motivated investors to make profit from offshore wind [2, 3].

Germany follows the UK with almost 8 GW offshore wind installation as of 2021 [1]. The country is also well-known with renewable energy deployments such as solar and onshore wind. Germany achieved high deployment of offshore wind due to national renewable energy targets such as 80% of electricity demand by renewables and high feed-in-tariff rates around 0.120 $/kWh [3, 4].

In Asia, China shows a dramatic increase on offshore wind investment. In 2021 alone, more than 10 GW of offshore wind installation was realized in China [1]. China owes this immense improvement to the national renewable energy targets [5, 6]. Other countries which have offshore wind installation in Asia are Taiwan, South Korea, Vietnam and Japan [1]. Furthermore, India has vast installation of onshore and a potential investor on offshore wind.

Table 1 summarizes countries national renewable energy targets and feed-in-tariff values.

Table 1. Comparison of countries by national renewable energy targets and feed-in-tariff values

Country	National REN target	Feed-in-Tariff
UK	50 GW offshore wind by 2030 [7]	0.246 $/kWh [3]
Germany	80% of the electricity by REN [4]	0.120 $/kWh [3]
China	1200 GW REN by 2030 [6]	0.134 $/kWh [6]

1.2 A Swot Analysis for Turkey

In this study, a Strengths, Weaknesses, Opportunities and Threats analysis is applied for an offshore wind investment in Turkey. The results are listed in Table 2 and are explained in the following subsections.

Table 2. SWOT analysis results for Turkey in terms of offshore wind investment

Internal Conditions	External Conditions
Strengths: • ***Four seas around the country*** • ***High wind speed on the seas*** • ***Developed know-how on wind energy***	Opportunities: • *Climate change actions that promote offshore wind* • *Close proximity to the leading offshore countries* • *Potential for leading neighbor countries with offshore wind expertise*
Weaknesses: • ***No strategic plan for offshore wind*** • ***Too many uncertainties with military and environmental constraints*** • ***Lack of infrastructure capability for a large deployment***	Threats: • *Long planning and permitting processes* • *High initial and operational cost of offshore wind* • *Intermittency and reliability problem of offshore wind*

1.3 Strengths

1.3.1 S1- Four Seas Around the Country

Turkey is surrounded by three seas, namely Black Sea (in the north), Aegean Sea (in the West) and Mediterranean Sea (in the south). Also, there is a significant internal Marmara Sea which connects Black Sea to Aegean Sea. Not many countries have shores to the seas which have constant and higher wind potential comparing to the lands. Therefore, Turkey has a strength of being surrounded by ready to be cultivated wind energy areas [8].

1.3.2 S2- High Wind Speed on the Seas

Wind speed on the seas are mostly more steady and higher than in land locations. However, some shores do not have high wind speed. Turkey's shores, especially Aegean shores have excessive wind potential. Global wind atlas shows that average wind speed at 100 m height reaches to 10 m/s on the northern Aegean seas where have higher wind potential than most of the countries' shores [4]. An offshore farm is mostly profitable on a site which has more than 7 m/s average wind speed [9].

1.3.3 S3- Developed Know-How on Wind Energy

Turkey has one of the highest installation onshore wind capacity in the World. As of 2021, the cumulative installation capacity from wind energy is around 11 GW which makes the country 12th biggest in terms of onshore installation [10]. This situation implies that Turkey has an accumulated know-how on the wind energy sector from installation to the operation and maintenance. There are many component manufacturers in the country such as tower, blade, generator, transformer and cables [11]. Therefore, Turkey has a potential of knowledge and techniques to be used in offshore wind energy sector.

1.4 Weaknesses

1.4.1 W1- Prolonged Planning and Permitting Processes

The planning and permitting processes of offshore deployment takes long time. For instance, the first offshore wind farm in the US, namely Block Island wind farm, was signed agreement between the national grid and the contractor in 2009, while the project commissioned in 2016 [12, 13]. China, on the other hand, had fast planning and permitting processes which could be good examples for Turkey [3].

1.4.2 W2- High Initial and Operational Cost of Offshore Wind

Turkey's primary energy depends on mostly imported natural gas which does not require a capital cost. Also natural gas is an inexpensive energy source. Therefore, Turkey's potential deployment of offshore wind is challenged by high initial cost of offshore wind. Moreover, comparing to onshore counterpart, offshore wind farm operations and maintenance cost is higher. This threat could only be tackled by policies and incentives which promote offshore wind deployment.

1.4.3 W3- Intermittency and Reliability Problem of Offshore Wind

The threat of intermittency and reliability of offshore wind is a general problem for all countries. Although, offshore wind has less intermittent nature due to steady and high winds with respect to onshore wind, lower availability is a potential threat for Turkey which is a country of having constant increase in energy demand.

1.5 Opportunities

1.5.1 O1- Climate Change Actions That Promote Offshore Wind

Climate change brings negative impacts to the humanity. To stop and eliminate the impacts of climate change, countries agreed on Sustainable Development Goals which include clean and affordable energy (Goal 7). This agreement encourages countries to invest more on renewable energy systems. Offshore wind is one of the most highly developing renewable energy systems due to this climate action plan which is an opportunity for countries such as Turkey.

1.5.2 O2- Close Proximity to the Leading Offshore Countries

Turkey has an opportunity to be at the close proximity to the leading offshore countries such as Germany, Netherlands, Denmark and UK. This is important because of the knowledge transfer on installation and operation and maintenance as well as financing from the experienced countries to Turkey which has no experience so far on the offshore wind energy.

1.5.3 O3- Potential for Becoming A Leading Country to the Neighbor Countries

Another opportunity for Turkey is to be at the close proximity of the other countries which have high offshore wind potential and no offshore installation such as Azerbaijan, Russia and Ukraine. Turkey's experience with onshore wind and potential investments on offshore could enable the country to be a leader country for the others which could seek assistance for an offshore deployment.

1.6 Threats

1.6.1 T1- No Strategic Plan for Offshore Wind

There was a tender for the first offshore wind farm in Turkey in 2018, however there was no bids at the end of the tender [8]. Therefore, the first attempt was unsuccessful to make an offshore wind initiation. This unsuccessful attempt was due to lack of detailed technical analysis, sufficient incentives and relatively higher cost of such an investment at the time of the tender. Although there are abundant areas on the seas, greater wind speed and know-how potential in Turkey for an offshore wind investment, there is still no concrete strategic plan or policies for one in the country.

1.6.2 T2- Too Many Uncertainties with Military and Environmental Constraints

Marmara Sea is protected by a recent regulation which bans any construction on the Sea due to mucilage occurrence [14]. Moreover, on the Northern Aegean Sea there are many military areas which are prohibited to do any actions [15]. Furthermore, alongside the Aegean Sea there is a short distance to shore due to the maritime borders of the country which limits the deployments of the offshore wind. On top of these constraints, there are uncertainties of potential archeological relics can be counted as weaknesses of investing on offshore wind on Turkey's seas [16].

1.6.3 T3- Lack of Infrastructure Capability for Large Deployments

Offshore wind energy requires additional infrastructure systems comparing to the onshore counterparts such as ports, offshore substations and underground cables. In Turkey, there are more than seventy ports, however, none of them was designed and specialized for an offshore wind installation or maintenance activities [9].

2 Conclusions

This paper summarizes the developmental reasons of the offshore wind in Europe and Asia, namely UK, Germany and Asia. Also, a SWOT analysis is applied for Turkey, which is a country of increasing energy needs and energy dependent. The main conclusion from this paper is that achiever countries succeeded due to mainly national targets and significant incentives for offshore wind and this could be an important lesson for Turkey. Finally, although Turkey is relatively late to the offshore deployment, the country has a potential to overcome the challenges and use its opportunities of high winds from their seas.

References

1. GWEC. Global Wind Energy Report (2022)
2. Yoshiaka, N.: The UK's Approach to Climate Change as Seen in Offshore Wind Power Generation. OPRI Perspectives (2020)
3. OECD. Renewable energy feed-in tariffs (2022). https://stats.oecd.org/Index.aspx?DataSetCode=RE_FIT
4. Whitlock, R.: Germany approves largest renewable energy expansion plan in history (2022). https://www.renewableenergymagazine.com/panorama/germany-approves-largest-renewable-energy-expansion-plan-20220708
5. Fan, Q., Xin, W., Jing, Y., Xin, L., Hao, H., Peng, L.: A review of the development of key technologies for offshore wind power in China. J. Marine Sci. Eng.**10**(7), 929 (2022). https://doi.org/10.3390/jmse10070929
6. Göss, S.: China should comfortably meet its 2030 Renewables target. But its emissions? (2022). https://energypost.eu/china-should-comfortably-meet-its-2030-renewables-target-but-its-emissions/
7. Major acceleration of homegrown power in Britain's plan for greater energy independence (2022). https://www.gov.uk/government/news/major-acceleration-of-homegrown-power-in-britains-plan-for-greater-energy-independence
8. Global Wind Atlas. https://globalwindatlas.info/
9. Karipoğlu, F. , Öztürk, S., Genç, M.S.: Determining suitable regions for potential offshore wind farms in Bandırma Bay using multi-criteria-decision-making method. MühendislikBilimleriveAraştırmalarıDergisi **3** (1), 123–132 (2021). https://doi.org/10.46387/bjesr.900204
10. TUREB. Turkish Wind Energy Statistics (7 Oct. 2022). https://www.tureb.com.tr/
11. Ministry of Foreign Affairs of Denmark. Wind Energy Market in Turkey (2021)
12. Reuters. Deepwater to build Rhode Island offshore wind farm (2008). https://www.reuters.com/article/utilities-rhodeisland-wind-idUSN2553068420080925
13. Dong, L., Lang, C.: Do views of offshore wind energy detract? A hedonic price analysis of the Block Island wind farm in Rhode Island, Energy Policy, vol. 167 (2022)

14. Daily Sabah. Turkey's Marmara Sea receives new environmental protection status (2021). https://www.dailysabah.com/turkey/turkeys-marmara-sea-receives-new-environmental-protection-status/news
15. SHODB. Denizcilereilanlaryıllığı (2021). https://www.shodb.gov.tr/shodb_esas/orj/2021_YILIK_e5be1.pdf
16. Dönmez, G.m Kaya, H.: Pre-historical submerged settlements in the sea of Marmara Coasts. Int. J. Environ. Geoinform. **7** (3), 335–346 (2020). https://doi.org/10.30897/ijegeo.788955

The Combined Effects of Channel Amplitude and Fluid Elasticity on Viscoelastic Fluid Flow Through a Periodic Channel

M. Madi[1](✉), A. Rafiki[2], K. Souhar[1,3], and Y. Haddout[4]

[1] Laboratory of Energy Engineering Materials and Systems, ENSA, Ibn Zohr University, Agadir, Morocco
madi.mohamed122@gmail.com

[2] Research Team ERME, Polydisciplinary Faculty of Ouarzazate, Ibn Zohr University, Box 638, Agadir, Morocco

[3] High School of Technology, Hassan II University, Casablanca, Morocco

[4] Research Team ERMAM, Polydisciplinary Faculty of Ouarzazate, Ibn Zohr University, Ouarzazate, Morocco

Abstract. Stability of second order and second grade viscoelastic fluid flow in a periodic channel in the inertial regime is studied. We examine, by using the local linear stability analysis, the combined effects of the periodic channel and fluid elasticity on the flow regime. A derived fourth order dispersion equation is developed and resolved by using the spectral method to analyze the effects of the governing parameters. The critical Reynolds number (Re_c) and wave number (α_c) values are established at various values of an elastic number (E), depending on the channel magnitude (ϵ), in accordance with the channel cross section (x). The found results show the presence of important stations: the enlarged and narrowed zones, the latter being the section most influenced on the flow, that corresponds to the minimum Re_c.

Keywords: Viscoelastic fluid · Periodic channel · Local linear stability analysis · Spectral method

1 Introduction

Due to the diversity application of the viscoelastic fluids flow in industrial processes, the research on the transition from a laminar to a turbulent flow regime has attracted much attention with the prospect of an appreciable decrease in turbulent friction in polymeric flows [1].

This illustrates the need to understand viscoelastic flows can be encountered in different cases, such as many disciplines of engineering research and technology, including oil recovery methods in industry and modeling of blood flow in Hemodynamics. This has motivated researchers to investigate viscoelastic flows using model geometries like the periodic channel, which capture the state of the environment and represent a class of mixed-kinematic flows [2].

S. Haddout et al. (Eds.): CCORE 2022, SPEES, pp. 72–78, 2023.
https://doi.org/10.1007/978-3-031-26967-7_6

The second order model presented by Rivlin and Ericksen in 1955 [3]. They were selected as a departure base for examining the stability behavior of viscoelastic fluids. It is a member of a broader category of fluids known as order models. It is a part of the family of constitutive equations [4], which was one of the first kind to be introduced as a way to represent deviations from the norm and Non-Newtonian behavior, these models can reflect non-null normal stress variations, which is a key property of viscoelastic fluids [5].

In this paper, we investigate the theory of the local linear hydrodynamic stability of a periodically channelized flow of a second order fluid (SO) for which (E < 0) and second grade fluid (SG) for which (E > 0). Our investigation focuses on the combined impact of the periodical channelling and the elasticity of the fluid on the stability of this flow. We examine the effect of geometrical and rheological parameters using the Chebyshev spectral collocation method based on collocation points of Gauss-Lobatto to resolve the eigenvalue problem [6].

2 Formulation of the Problem

In the dimensionless Cartesian coordinate system (x, y), we explore a plane Poiseuille flow of an incompressible viscoelastic fluid on two periodic walls defined by:

$$y_w = \pm(1 + \epsilon sin(nx)), \tag{1}$$

where ϵ is the channel amplitude, and n is the wavenumber of the walls.
The geometry of the physical issue examined for a periodic domain is depicted in the Fig. 1.

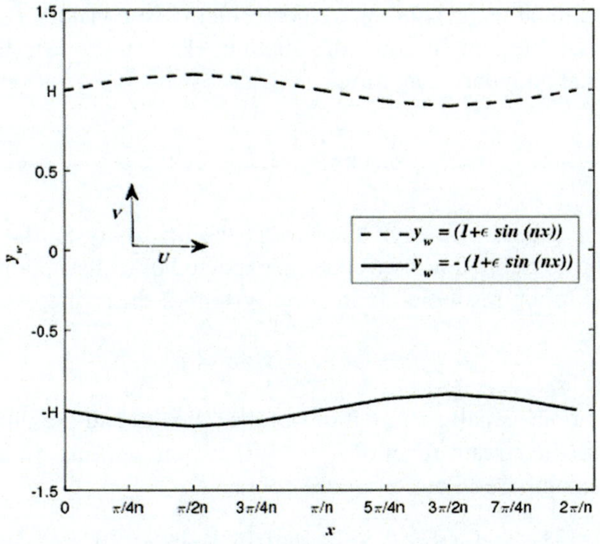

Fig. 1. Schematic representation in Cartesian coordinates.

The equations governing the motion of the fluid in dimensionless shape are written as:

$$Re(\frac{\partial u}{\partial t} + u.\nabla u) = -\nabla p + \nabla.(\tau), \tag{2}$$

$$\nabla.u = 0, \tag{3}$$

where P is the dimensionless pressure, $u = (U, V)$ is the dimensionless velocity field and τ is the stress tensor defined by the following law of behavior:

$$\tau = \frac{1}{Re}A_1 + E_1A_2 + E_2A_1^2, \tag{4}$$

The corresponding boundary conditions are defined by:

$$u(x, y = \pm y_w) = 0. \tag{5}$$

The control parameters are $Re = \frac{\rho u_0 H}{\mu}$ and $E_i = \frac{\gamma_i}{\rho H^2}$ denote the Reynolds number and the elasticity number, respectively.
In these relations, ρ is the density of the fluid, μ is the dynamic viscosity, H is the average channel length, u_0 is the velocity at the center of the channel, γ_1 and γ_2 are material constants of the fluid representing elasticity and the cross viscosity, respectively.
The quantities A_1 and A_2 are the kinematic tensor systems, with $A_1 = (\nabla u) + (\nabla u)^{\mathrm{T}}$, and with $\mathrm{A}_2 = \frac{\partial_j \mathrm{A}_1}{\partial t} = \frac{\mathrm{DA}_1}{\mathrm{Dt}} + \mathrm{A}_1.(\nabla \mathrm{u}) + (\nabla \mathrm{u})^{\mathrm{T}}.\mathrm{A}_1$. Where A_1 is the strain rate tensor, $\frac{\partial_j}{\partial t}$ is the Jaumann derivative and $\frac{D.}{Dt}$ is the material derivative.

3 Linear Stability Analysis and Numerical Solutions

Developing the fundamental solution to the problem, we consider a stationary and unidirectional flows, taking into account of equations (Eq. 2), (Eq. 3), (Eq. 4), the solution which satisfies the boundary condition is:

$$u_b(y) = \frac{1}{y_w}\left(1 - \frac{y^2}{y_w^2}\right) \tag{6}$$

According to the ordinary rules of the linear stability analysis, the perturbed flow is displayed as a basic stable flow as well as an expected time disruption infinitesimal. We disrupt the equilibrium resolution to find the system of the shape:

$$u = u_b + u'; v = v'; p = p_b + p', \tag{7}$$

where u', v', p' are the small perturbations of the velocity and pressure component.
It is usual to use the stream function $\psi(x, y, t)$, when working on a two-dimensional problem. The decomposed Fourier modes defined by:

$$[\psi](x, y, t) = [\varphi](y)\exp^{[i\alpha(x-ct)]}, \tag{8}$$

where $\mathrm{i} = \sqrt{-1}$, α is the wave numbers and $\varphi(y)$ the complex magnitude of the perturbation ψ (x, y, t), and $\boldsymbol{c} = \boldsymbol{c_r} + \mathrm{i}\boldsymbol{c_i}$ is the complex wave speed.

Taking into account the equations (Eq. 5), (Eq. 6) and (Eq. 7), the dispersion equation called modified Orr-Sommerfeld is written as:

$$\begin{bmatrix} \left[(1 + i\alpha ERe u_b)\left(D^2 - y_w^2\alpha^2\right)^2\right] - \\ i\alpha y_w^2 Re\left[u_b\left(D^2 - y_w^2\alpha^2\right) - u_b D^2\right] + \\ i\alpha ERe u_b D^4 \end{bmatrix}\varphi = \begin{bmatrix} i\alpha EcRe\left(D^2 - y_w^2\alpha^2\right)^2 - \\ i\alpha c y_w^2 Re\left(D^2 - \alpha^2\right) \end{bmatrix}\varphi, \quad (9)$$

where $\frac{\partial^m}{\partial y^m}$ is a differential operator that is replaced by D^m, with $m = 1, 2$.
The boundary conditions for the numerical solution of the system of dispersion Eq. (9) are defined as follows:

$$\phi(\pm 1) = D^1\phi(\pm 1) = 0. \quad (10)$$

The equations and the associated boundary conditions describe completely the evolution of an infinitesimal perturbation in time and space. It consists reducing an initial value problem into an eigenvalue problem, such that (Eq. 9) is written as:

$$A.f = cB.f, \quad (11)$$

where A, B two operators depending on Re, α and c.
The eigenvalue problem (Eq. 11) is calculated numerically employing the spectral collocation method established on Chebyshev polynomials estimated in N Gauss-Lobatto collocation points.
In the interest of verifying our operating code, we have calculated Re_c for a Poiseuille plane flow using a Newtonian fluid (E = 0, ε = 0). We discover that Re_c is equal to 5772.22 for α_c=1.02, at N = 60. In addition, in Fig. 2, we trace the eigenvalue spectrum $\sigma_i = f(\sigma_r)$ at Re = 2013, α = 1.3, ε = 0 and N = 120, with $\sigma = \sigma_r + i\sigma_i$ is the complex wave frequency.
This appearance is nearly the exact same as the one mentioned above in Fig. 9 of Sureshkumar [3]. Therefore, in this analysis, we rely on N = 120 operating proposals to achieve performance that does not depend on the number of functions.

4 Results and Discussion

After validation of our numeric code, we introduce the stability analysis results describing the associated impacts of (*E*) and the channel's periodic modulation on stability.
We introduce in Fig. 3, the stability limits in the various sections of the channel corresponding to the wall magnitudes ε = 0.01 and ε = 0.03 for the three fluids: Newtonian (E = 0), SO (E < 0) and SG (E > 0). This Fig. 3 shows that, moving from the relaxed zone ***x***= 0 to the enlarged section ***x***= $\pi/2n$, the Re_c increases and the flow gets more stable. From section ***x***= π/2*n* to the narrower section ***x***= $3\pi/2n$, the flow's stability reduces, and then it finds the stability at section ***x***= $\pi/2n$ for various values of E. Moreover, It can be detected that, for all fluids, the critical wave numbers α_c decreases when Re_c increases and vice-versa.

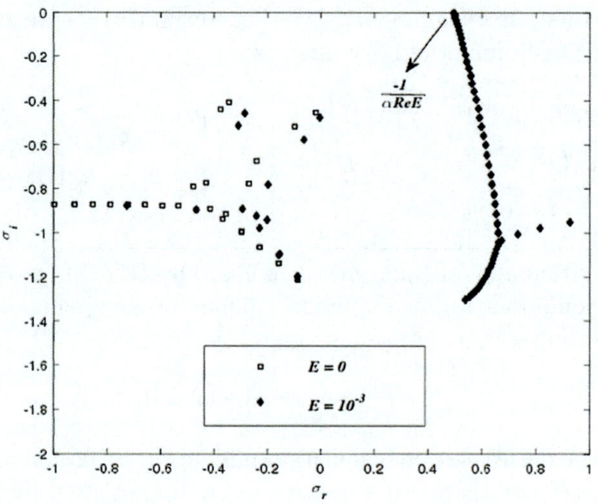

Fig. 2. The Newtonian and SO fluids eigenspectrums at $Re = 2013$, $\alpha = 1.3$ for $N = 120$, $\varepsilon = 0$.

As a result, the most significant station and the most instable flow region is $\boldsymbol{x_c}= 3\pi/2n$, which corresponds to the narrowest section when Re_c is minimal. When we compare the instability of three types of fluids, we note that the SG fluid is more stable than the Newtonian fluid which is more stable than SO fluid and this for different sections.

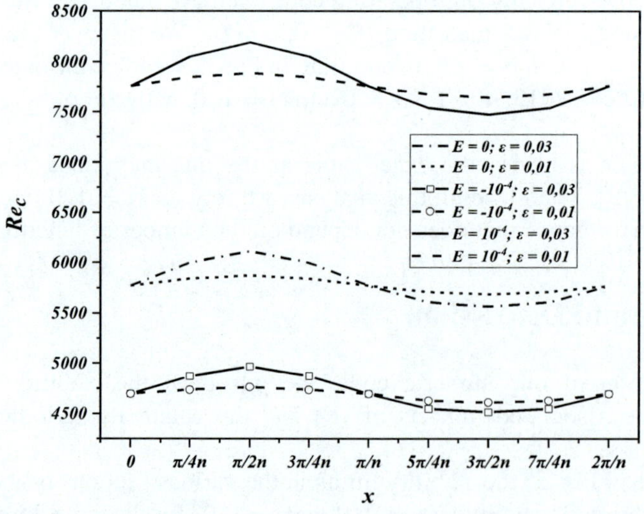

Fig. 3. The variation of Re_c as a function of different cross sections x.

In Fig. 4, we introduce the evolution of Re_c according to channel magnitude ε. It can be seen that when ε increases, Re_c increases in the enlarged section ($x_c = \pi/2n$) and reduces in the restricted cross section ($x_c = 3\pi/2n$). This result allows to deduce that

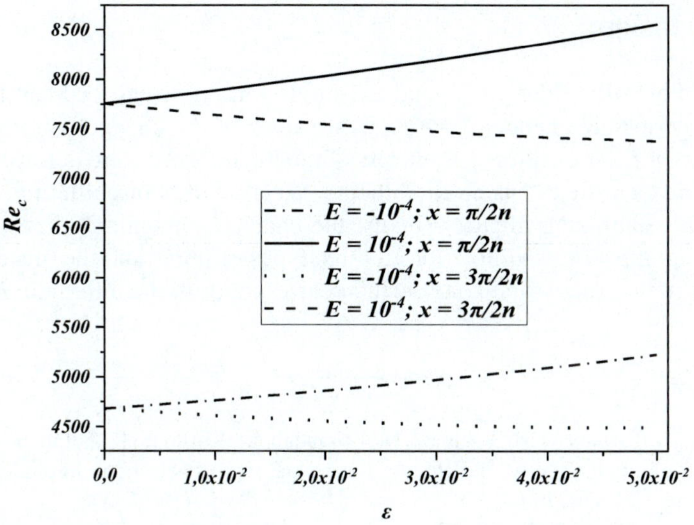

Fig. 4. The evolution of Re_C versus channel magnitude ε.

the channel magnitude has a stabilizing or destabilizing role of the flow in relation to the channel section for the three categories of fluids (Newtonian, SO and SG).

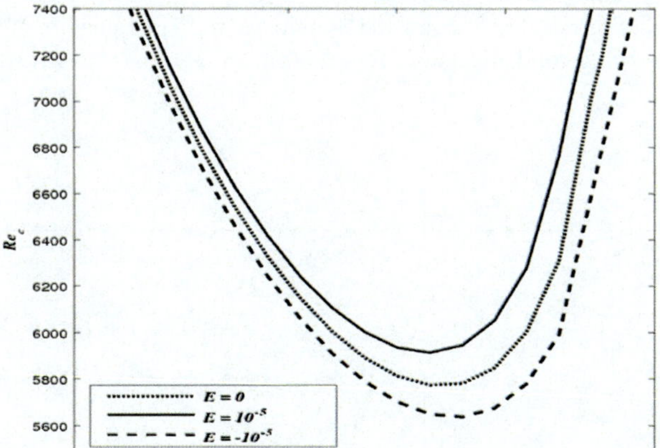

Fig. 5. The neutral stability curves corresponding to several values of the elasticity E at $\varepsilon = 0$ for $x = 0$.

In Fig. 5, we provide the neutral stability curves of three types of fluid flow, we note that the second grade (SG) is more stable than Newtonian fluid. While the second order (SO) is more unstable.

5 Conclusion

The plane Poiseuille flow of SO and SG fluids was surveyed for local linear stability through a periodical channel, for specified values of the channel amplitude ε and the elastic number E. The critical parameters of the Re, α_c and c were defined. We conclude that SO fluid's elasticity has a destabilizing effect on the Poiseuille flow, while that of SG fluid has stabilizing impact. Finally, the channel's magnitude has a stabilizing or destabilizing impact on the flow for all three types of fluids and the presence of a more significant station (narrowed zone), which corresponds to the minimum Re_c value.

References

1. Sadeghy, K., Taghavi, S.M., Khabazi, N., Mirzadeh, M. Karim fazli., I.: On the use of hydrodynamic instability test as an efficient tool for evaluating viscoelastic fluid models. Adv. Studies Theor. Phys. **1**, 367–379 (2007)
2. Sureshkumar, R.: Local linear stability characteristics of viscoelastic periodic channel flow. J. Non-Newtonian Fluid Mech. **97**, 125–148 (2001)
3. Rivlin, R.S., Ericksen, J.L.: Stress deformation relations for isotropic materials. J. Rational Mech. Anal. **4**, 323–425 (1955)
4. Dunn, J.E., Rajagopal, K.R.: Fluids of differential type : critical review and thermodynamic analysis. Int. J. Eng. Sci. **33**, 689–729 (1995)
5. Cracco, M., Davies, C., Phillips., T.N.: Linear stability of the flow of a second order fluid past a wedge. Phys. Fluids **32**, 084102 (2020)
6. Trefethen, L.N., Embree, M.: Spectra and Pseudospectra: The Behavior of Nonnormal Matrices and Operators. Princeton University Press (2005)

Impact of Climate Change on Water Resources Management

Evaluation of Floc Characteristics Induced by Heavy Metals in an Estuarine Environment

Febina A. Manaf[1,2(✉)], K. L. Priya[1], Hamie Harold[3], and Suchith Chellappan[3]

[1] Department of Civil Engineering, TKM College of Engineering, Kollam, India
[2] APJ Abdul Kalam Technological University, Kerala, India
febina.manaf@gmail.com
[3] Department of Civil Engineering, UKF College of Engineering and Technology, Kollam, India

Abstract. Experimental investigations were carried out to understand the role of metal fraction and turbulence on the flocculation of kaolin flocs. Mixing experiments were done in a flocculatorand themicro-scale investigations were carried out using an image-capturing system followed by an image-processing technique. The floc characteristics were analyzed for different meal concentrations (0, 0.1, 0.5, 1, 5, 10, 20 mg/l), salinities (0, 15 and 30 g/L), turbulent conditions (0, 5, 10, 20, 30 and 40 s^{-1}) and time intervals (0, 10, 20 and 30 min). The work was followed by an analysis of floc size, fractal dimension, floc density, and floc volume fraction in three size classes: 0–50 μm, 50–100 μm, and >100 μm. The maximum floc size was observed at a salinity of 30 g/L and turbulence shear of 10 $s^{-1}$for the highest heavy metal concentration of 20 mg/l. An increase in heavy metal concentration enhanced the aggregation process at high salinity conditions and low turbulence shear rates. The rise in turbulence shear initially increased floc size, which further decreased at high turbulence shears. Resistance of flocs to breakage was analyzed using the parameter breakage coefficient. The breakage coefficient is the ratio of the number of flocs before breakage to the number of flocs after breakage.The breakage coefficient of macro-flocs was determined from the experimental results and a relationship in terms of metal concentration is proposed. The study suggests that the binary breakage model finds application at intermediate turbulence ranges of 20 s^{-1} with the highest metal concentration of 20 mg/l. Thus, a relation for the breakage coefficient in terms of metal concentration was generated through the detailed analysis of the breakage coefficient from the experimental outcomes. Therefore, the developed model can be applied to quantify the breakage coefficient under all ranges of turbulence shear encountered in estuaries with mixed sediments.

Keywords: Floc characteristics · Metal concentration · Salinity · Turbulence · Breakage coefficient

1 Introduction

Estuaries are semi-closed coastal water bodies where freshwater coming from the land dilutes seawater and where there is unrestricted water exchange with the sea. Estuaries act as transitional zones between habitats in rivers and those in the ocean. In the estuaries,

S. Haddout et al. (Eds.): CCORE 2022, SPEES, pp. 81–101, 2023.
https://doi.org/10.1007/978-3-031-26967-7_7

a number of physical, chemical, and biological processes take effect, however, flocculation is the one that results in the settlement and removal of suspended sediments from the estuary.Flocculation processes are aided by the influence of turbulence shear, suspended sediment concentration, presence of organic matters, toxic heavy metals, salinity, pH, viscosity, etc. the flocs within the water column undergoes the processes of transportation, flocculation, settling, deposition, erosion and resuspension [8]. Salinity, suspended sediment concentration, turbulence, particle size, and particle shape, are some of the additional parameters that influence the flocculation and settlement of clay particles. Hydrogen bonds or Vander Waals forces helps to thefloc generation through aggregation. In aqueous environments, most particles are charged as a result of both surface ionisation and specific ion adsorption. The distribution of ions on the surface is influenced by the surface charge [6]. Attraction of oppositely charged particles and repulsion of equally charged particles, along with the tendency for thermal motion, as well as the potential for ionic attraction or repulsion, results in the developmentof an electrical double layer. The physical mechanisms influencing flocculation include Brownian motion, turbulent shear, and differential settling. Brownian motion plays a significant role only if the particle size is less than 1 μm [10], and the suspended sediment concentration (SSC) is much higher than 10 g/L and therefore can be neglected in the natural environment. The concentration of suspended sediments, the composition of suspended sediments, salinity, pH, temperature, and viscosity are the main chemical parameters that influence the flocculation of cohesive sediments. The flocculation process can be disrupted by high levels of turbulence that happen throughout a tidal cycle because they cause floc break-up. The simultaneous impact of turbulence on aggregation and floc-break-up was investigated by [1, 10]. Salinity has been well-studied to cause the rate of flocculation to increase. Salinity gradient plays a major role in estuarine sediment dynamics [12]. Salinity imparts some chemical changes in estuaries which contribute to the heavy metal removal in estuaries [2]. Salinity-related changes include a decrease in the concentration gradient between charged sediment's Stern and Guoy layers as well as the thickness of the diffuse double layer. Then, the repellent force between two charged particles exists only close to the particle and is overpowered by Van der Waals attraction, which results in coagulation. Increased double-layer thickness results in decreased cohesiveness and aggregation when pH levels rise. The distribution of heavy metals has been impacted by suspended sediment resuspension and movement [9]. The phase change of heavy metals, which may be of natural or anthropogenic origin [3, 7] has a significant impact on their fate and transportation. In polluted estuaries, external metal sources might be significant. In rivers, Fe is primarily present as particulate (including colloidal) Fe(III) oxide, which is its most stable form in air-saturated waters at close to neutral pH levels. A significant amount of particulate Fe may be deposited in estuarine sediments, yet some may still be transported through estuaries to the ocean. Studies have been reported on the flocculation behavior of Fe(II) in the water treatment processes. Since Fe(II) is a major heavy metal present in almost all estuaries [2] the flocculation of the same will be predominant in estuaries also, which is yet to be studied. Enhanced flocculation imparted by metals was observed at higher salinity reaches [4] and pH [2]. A higher removal rate of heavy metals was reported during estuarine mixing. The present study examines how clay in an estuarine environment behaves in terms of flocculation under the impact of Fe(II) at five distinct turbulences and three salinity conditions. A relationship for breakage coefficient in terms of metal concentration at different turbulent conditions is proposed. The variation in floc volume fraction of different size classes for different time intervals under different conditions of metal concentration, salinity, and turbulences were also investigated to understand the settling behavior of kaolin flocs.

2 Materials and Methods

2.1 Instrumentation

The instruments used for the study include a jar test apparatus, an image capturing device, and an image processing unit which are described in subsequent sections.

2.1.1 Jar Test Apparatus

The experimental setup adopted for the study included a jar test apparatus equipped with a variable speed agitator; it also consisted of a single rectangular paddle of 5 cm width, placed 30 mm above the suspension containing 1 L glass beaker. The paddle is operated between 0–300 revolutions per minute (rpm). The turbulence shear (root mean square velocity gradient) G (s^{-1}) was obtainedfrom the power dissipation of the propeller, P [8]

$$G = \sqrt{P/\mu V} \tag{1}$$

where μ is the viscosity of the fluid, V is the volume of suspension, N_p is the power number

$$P = N_p \rho_w n^3 d^5 \tag{2}$$

where ρ_w is the density of the fluid, n represents the stirring frequency in rotations per second, and d is the diameter of the propeller. The turbulent shear adopted for the study ranged between 5–40 s^{-1} (5 s^{-1}, 10 s^{-1}, 20 s^{-1}, 30 s^{-1}, 40 s^{-1}).

2.1.2 Image Capturing System

The image capturing system includes a DSLR camera (Canon 800D) by using a reverse ring, the normal 18–55 mm lens of the camera is converted into a macro lens with 1 to 4 times magnification and restricting the lens's ability to change the aperture. The shutter speed and International Organisation for the Standardisation (ISO) have to be set as 1/1000,100 respectively after an optimized number of trials. For capturing the image the sample has to be placed in a rectangular glass cuvette of 1.5 cm square cross section and 4 cm height, placed in front of an LED light of 48 W.

2.1.3 Image Processing

Image processing is required to determine the number of particles and the size of the flocs from the raw image [14]. The captured images were processed using Image J software.

The different stages involved in image processing include (i) set scale image calibration; (ii) converting an image from RGB to 8-bit greyscale; (iii) Fourier filtering process; (iv) thresholding; (v) computation of floc characteristics. The image is calibrated against known values followed by applying to an uncalibrated image such that both images have the same magnification. Intensities of the Red, Green and Blue hues range from the darkest to the lightest in RGB images, which are composites of three independent images. The FFT > Bandpass filter step in Image J implements two Fourier filtering functions. Filtering an image for certain feature sizes by choosing the lowest and maximum feature size followed by cutting away a zero frequency stripe in the orthogonal direction in the frequency space to remove repetitive horizontal and vertical stripes. Image thresholding was done to separate an image's foreground from its background. Otsu thresholding was the auto threshold system that was used for solving the issue that arises when the input conditions for the software change. The final results give the total number of particles, area of the particle, major axis, minor axis, circularity, aspect ratio, roundness, solidity, etc., and the diameter of the floc was obtained. The fractal dimension, floc volume fraction, and floc density were quantified.

2.2 Experimental Procedure

In the present study, 1 g/L kaolin suspensions were used to represent cohesive sediments in an estuarine environment. For the analysis, a Jar Test apparatus and image capturing system was used. The salt of Iron ($FeSO_4$) was introduced into the suspensions at different concentrations of 0, 0.1, 0.5, 1.5, 10, and 20 mg/L. The suspensions were given initial turbulence of 20 s^{-1} rpm for 2 min for initial mixing up. Then the suspensions are

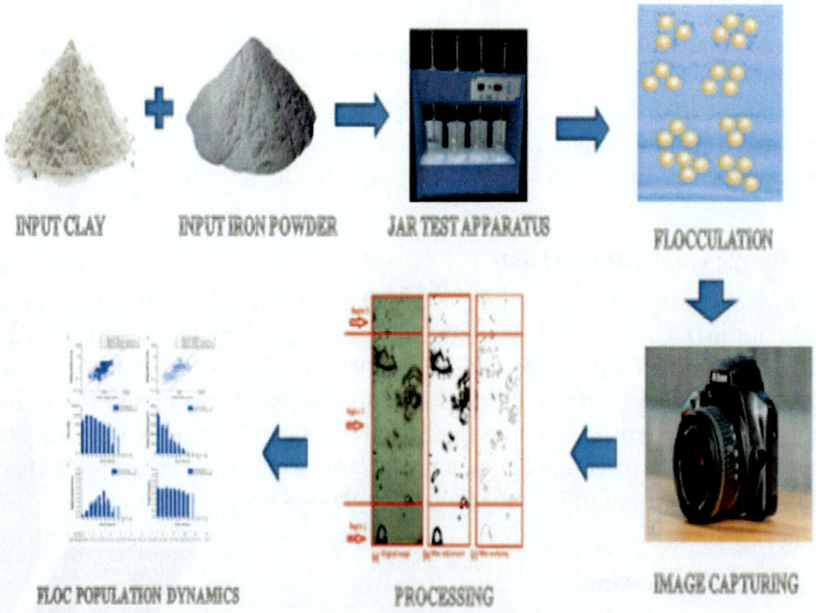

Fig. 1. Experimental procedure adopted for the study

provided with different turbulence of 5 s^{-1}, 10 s^{-1}, 20 s^{-1}, 30 s^{-1}, and 40 s^{-1} to enable floc formation for 10 min, followed by sampling at 10, 20, 30, and 40 min using a pipette for image capturing. The images of the flocs were taken using a DSLR camera and the captured images were processed using Image J software. From the computed data, floc size, fractal dimension, floc density, breakage coefficient, and floc volume fraction were quantified using Microsoft Excel software (Fig. 1).

3 Results and Discussion

3.1 Variation of Floc Volume Fraction with Metal Concentration

Floc volume fraction (FVF), the ratio total volume of flocs in each class to the total volume of flocs in all the size classes gives a real picture of the volumetric distribution of flocs of different classes. Three size classes are adopted in the study to represent microflocs and macroflocs. Size class I shows the particles with a size between 0 μm and 50 μm, Class II represents the particle with a size in the range of 50 μm to 100 μm, and class III represents the particles of size greater than 100 μm. The variation of FVF with metal concentrations is analyzed for different salinities and turbulences. Figures 2, 3 and 4 represents the influence of Fe(II) on the floc volume fraction at different salinity conditions (0 g/L, 15 g/L, and 30 g/L).

Figures 2, 3 and 4 show that with an increase in metal concentration and salinity, floc formation takes place due to agglomeration, whereas with an increase in turbulence, the breakup of flocs is predominant.

(i) At 0 g/l Salinity

Figure 2 shows the variation of floc volume fraction for 0g/L salinity.

(ii) At 15 g/l Salinity

The variation of floc volume fraction for salinity 15 g/L for different turbulences is shown in Fig. 3.

(iii) At 30 g/l Salinity

Figure 4 shows the effect of metal concentration on the floc volume fraction for a salinity of 30 g/L. The effect of metal concentration and turbulence had a significant effect on the floc volume fraction of kaolin particles at 30 g/L salinity. Higher salinity represents the region closer to the estuarine mouth. Significant changes in the flocculation can occur within this region, where the turbulence shear is also higher. More bigger-sized macroflocs were observed for this salinity under lower turbulence and higher metal concentration.

Figure 2 shows the floc volume fraction of various size classes under the influence of heavy metal at different turbulence shear rates without the presence of salinity. Figure 2 depicts that the size class (III) increases with an increase in metal concentration. Aggregation of particles takes place under the influence of metal concentration, which is achieved by reducing the Vanderwalls forces of attraction between the particles. As the concentration of heavy metal increases, more amount of positively charged particles get attached around the surface of kaolin, which is a negatively charged particle. Increased heavy metal concentration produces metal ions that reduce the Vanderwall forces between the clay particles thereby attracting more clay particles to stick together [5]. This bonding being very strong imparts resistance of flocs to breakage forming macroflocs at higher metal concentrations.

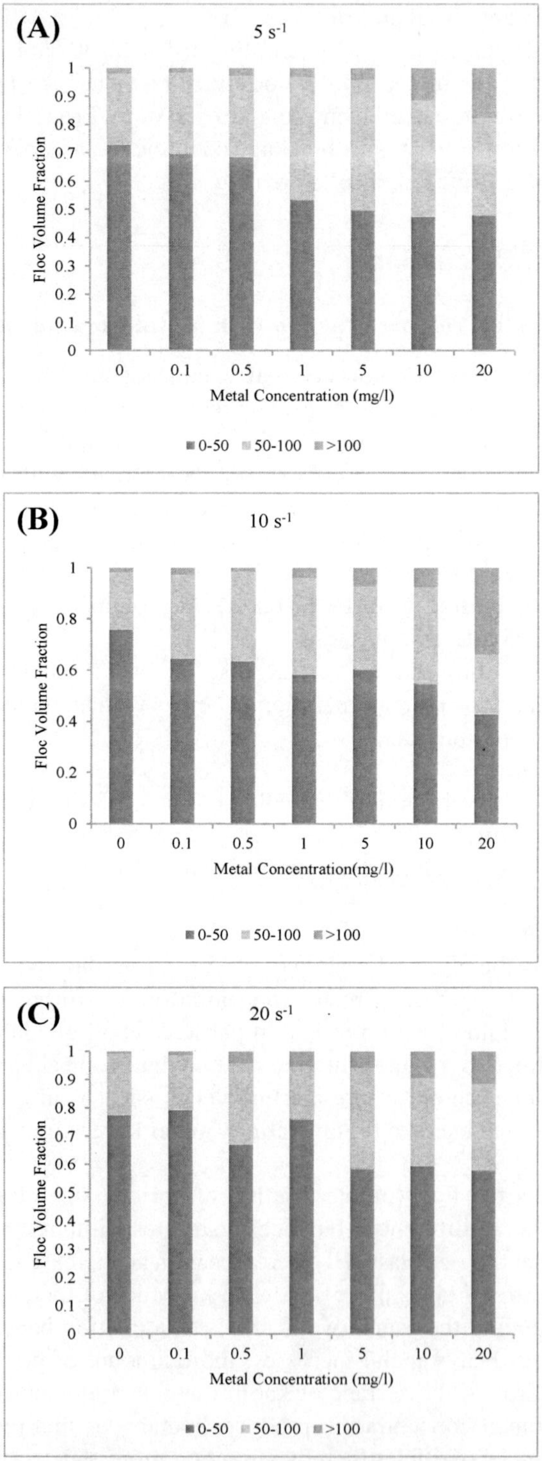

Fig. 2. Influence of heavy metal on the floc volume fraction at turbulences (A) 5 s^{-1} (B) 10 s^{-1} (C) 20 s^{-1} (D) 30 s^{-1} (E) 40 s^{-1}

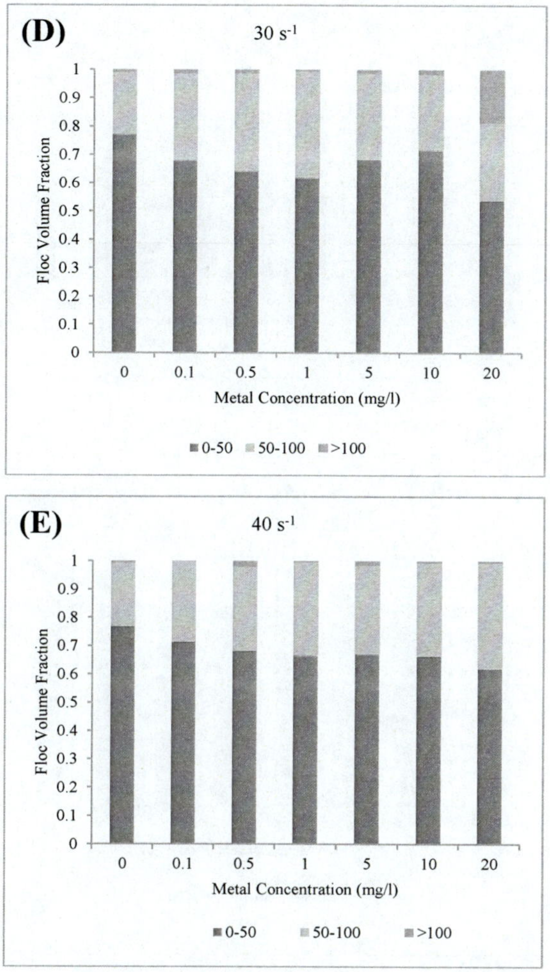

Fig. 2. (*continued*)

As the turbulence increases, the floc size initially increases at low turbulence and reaches optimum turbulence and a further increase in turbulence shear causes the disintegration of flocs. Thus, the maximum floc volume fraction that can be reached is at 10 s^{-1}. The microfloc population tends to increase with an increase in turbulence.

In Fig. 3, as salinity and metal concentration increases, particles tend to aggregate forming macroflocs. Salinity provides more surface charge to promote floc growth. Strong, big, and resistant flocs are formed in the presence of salinity [11, 17]. Since Fe(II) has a high coagulation potential, Fe(II) facilitates charge neutralization and promotes the aggregation of particles, producing more strong and more resistant flocs [4]. Thus the simultaneous effect of salinity and heavy metal concentration promotes flocculation. The macro flocs formation was observed to limit at high turbulences indicating the effect of shearing over the strong bonding between particles. Higher turbulences promote floc disruption [8].

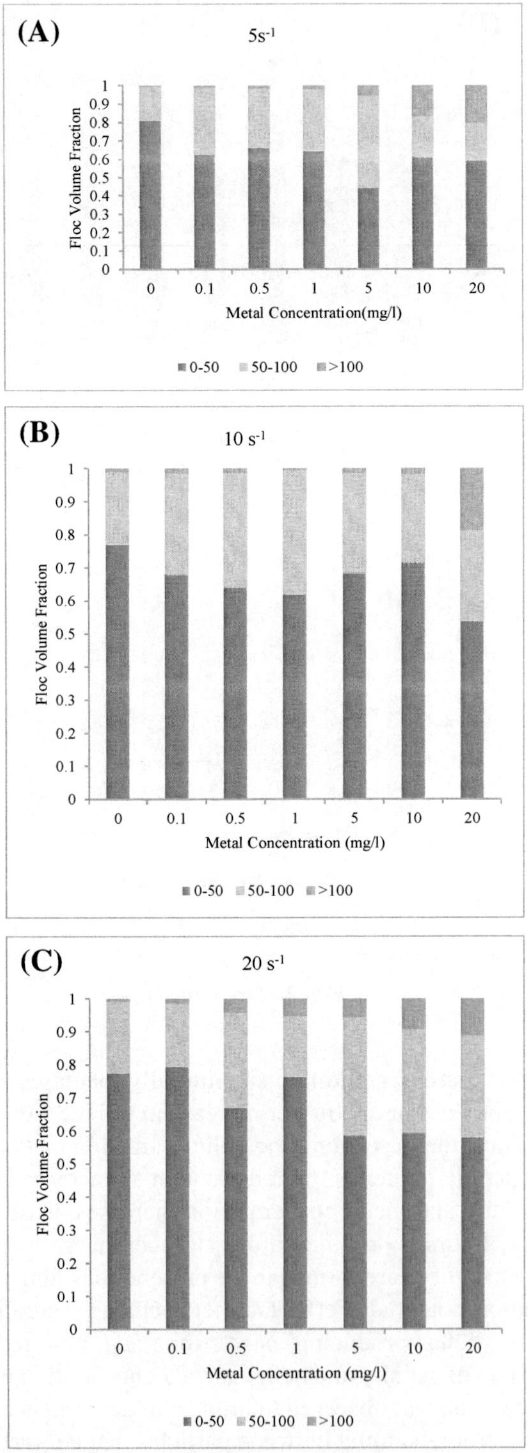

Fig. 3. Influence of heavy metal on the floc volume fraction at turbulences (A) 5 s^{-1} (B) 10 s^{-1} (C) 20 s^{-1} (D) 30 s^{-1} (E) 40 s^{-1}

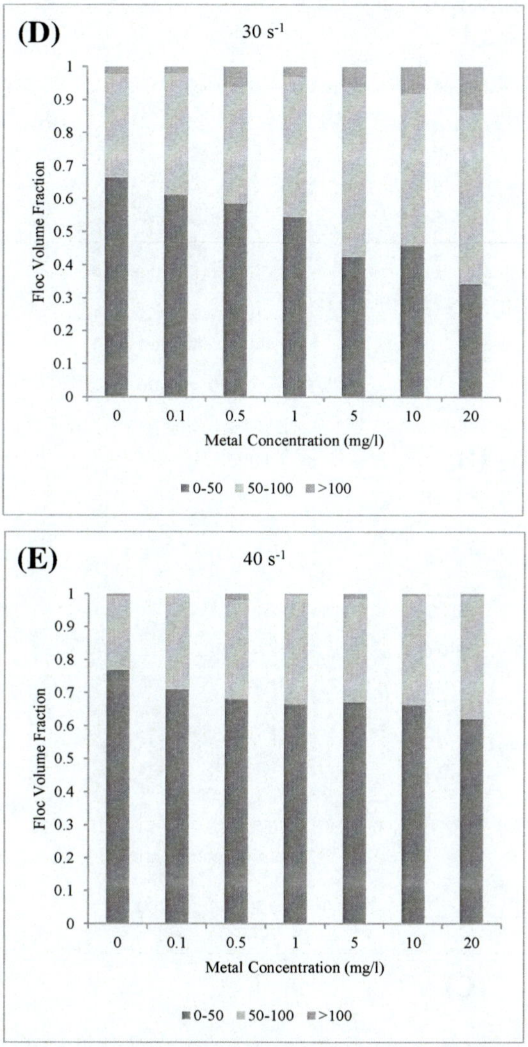

Fig. 3. *(continued)*

At salinity 30 g/l, the floc volume fraction of size class III exhibited the highest value for all turbulence shear (Fig. 4). The least population of microflocs was recorded at a salinity of 30 g/l. At high salinity, where more surface charges are available, macrofloc formation increases. An increase in metal concentration further adds to the floc formation by neutralizing the charges resulting in bigger-sized flocs [15, 16]. Kaolin in the presence of Fe produces stronger flocs with less restructure rate [21]. It was noticeable that the effect of turbulence was limited to the flocs formed at high salinity and metal concentration. Disruption of macroflocs (at 20 mg/L Fe(II) and 30 g/L salinity) was comparatively less than that for macroflocs at low salinity and metal concentration. The breakup of flocs at low salinity and metal concentration was predominant at higher shear rates.

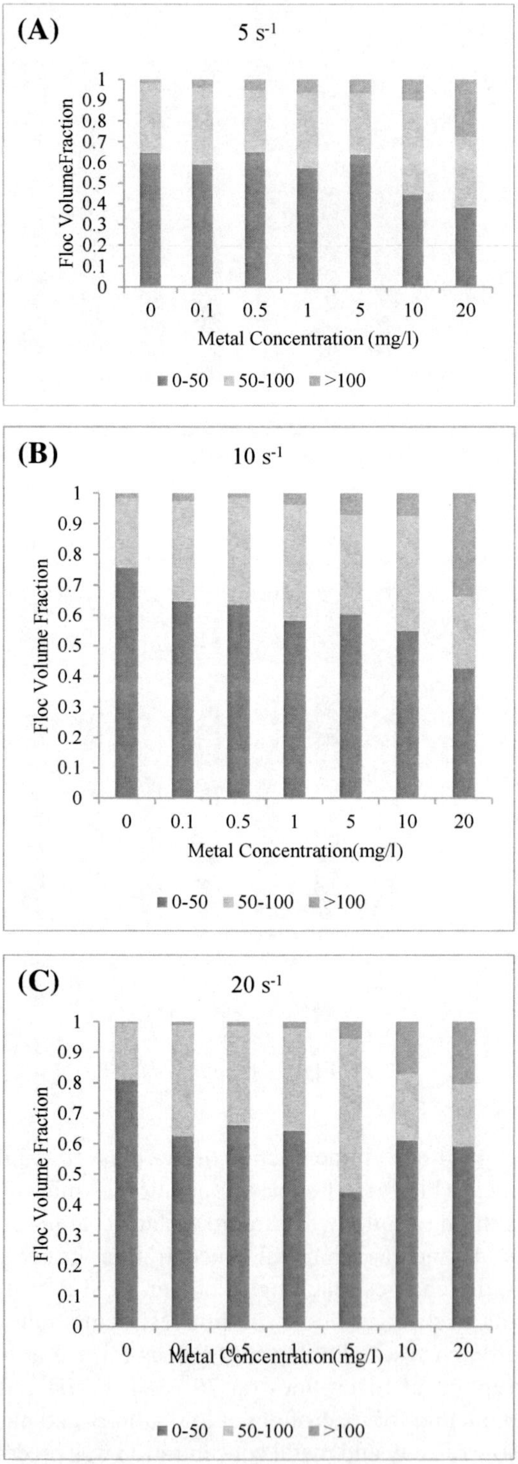

Fig. 4. Influence of heavy metal on the floc volume fraction at turbulences (A) 5 s^{-1} (B) 10 s^{-1} (C) 20 s^{-1} (D) 30 s^{-1}(E) 40 s^{-1}

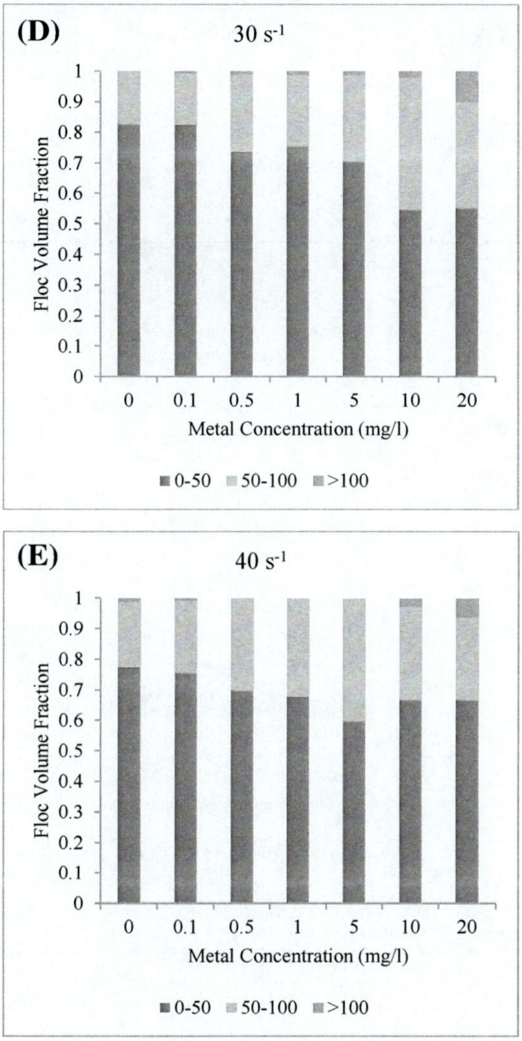

Fig. 4. (*continued*)

3.1.1 Variation of Maximum Floc Diameter with Heavy Metal Concentration

Figure 5 shows the variation of floc diameter with metal concentration for different salinities and various turbulent shears.

Figure 5 shows the influence of heavy metal on maximum floc diameter under different conditions of salinity and turbulence. With the increase in metal concentration from 0 g/L to 20 g/L at various salinities (from 0 g/L to 30 g/L), the floc size increases due to higher charge neutralization by Fe(II) and the presence of surface charges [5]. Maximum floc diameter is observed at salinity 30 g/l condition with the lowest turbulence rate of 10 s^{-1}. The increase in floc size is further restricted by the high turbulence shear. The floc size is highly skewed towards smaller size distribution under higher shear

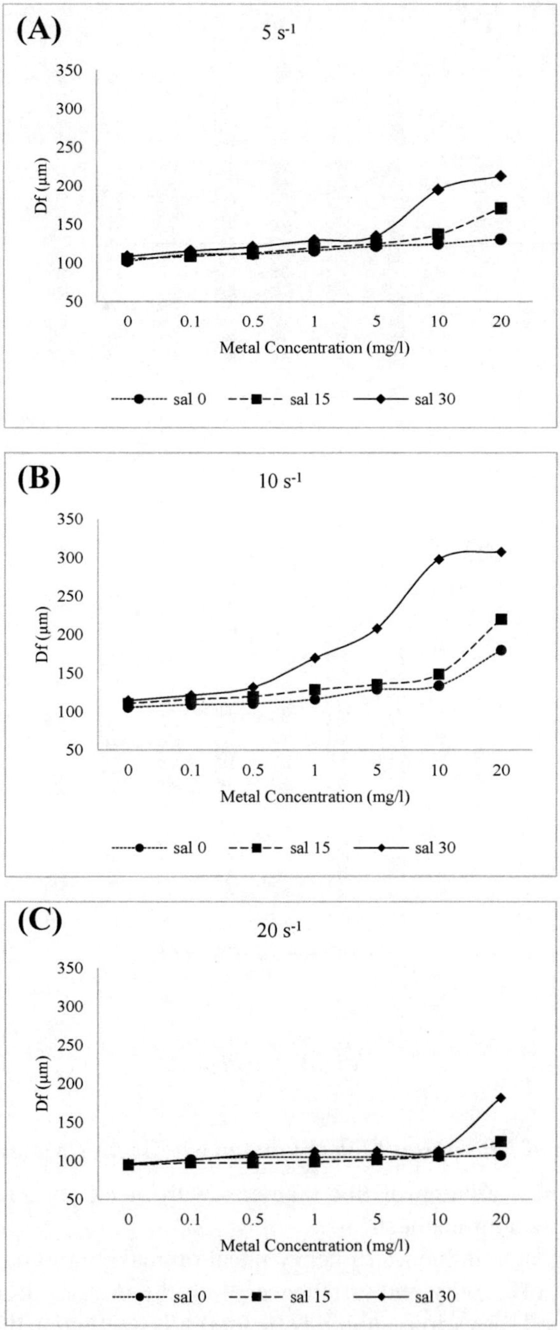

Fig. 5. Influence of heavy metal on D_{fmax} at turbulences (A) 5 s^{-1} (B) 10 s^{-1} (C) 20 s^{-1} (D) 30 s^{-1} (E) 40 s^{-1}

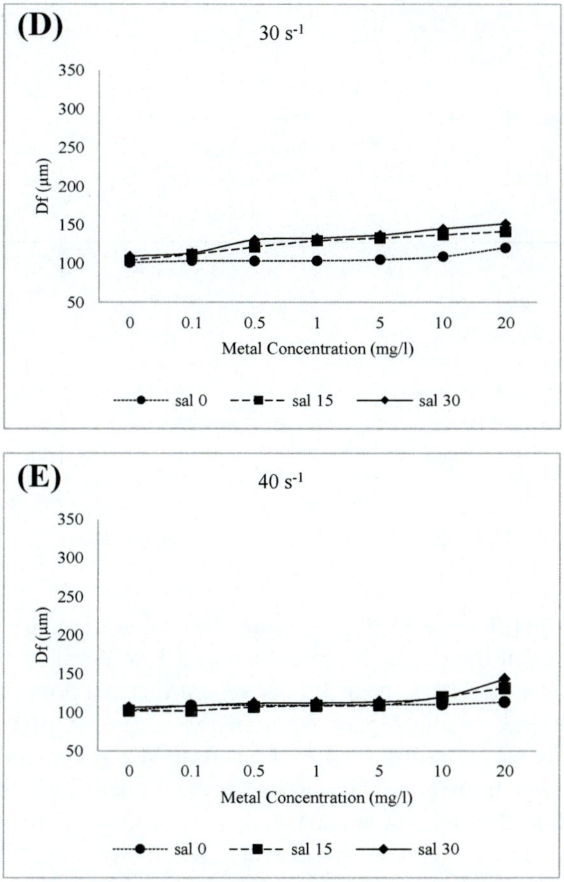

Fig. 5. (*continued*)

conditions. Therefore, it is demonstrated that the enhanced turbulence conditions cause the Df values to decrease [8]. This shows the formation of macroflocs occurs at lower shear rates, while higher turbulences result in their disintegration.

3.1.2 Variation of Floc Density with Metal Concentration

The influence of heavy metal on the floc density of kaolin with different turbulent shear rates at different salinity was analyzed. Floc density is represented as ρ_f

$$\rho_f = \left(D_p - 1\right)(2/D_f)^{(3-nf)} \tag{3}$$

where D_p = particle size of kaolin is 2.4, D_f = diameter of the floc, n_f = Fractal dimension

$$n_f = \log A/\log a \tag{4}$$

where A is the area of the floc, and a is the major axis length of the floc.

Figure 6 shows the variation of floc density with metal concentration for different turbulences. It is well understood that as metal concentration increases floc density

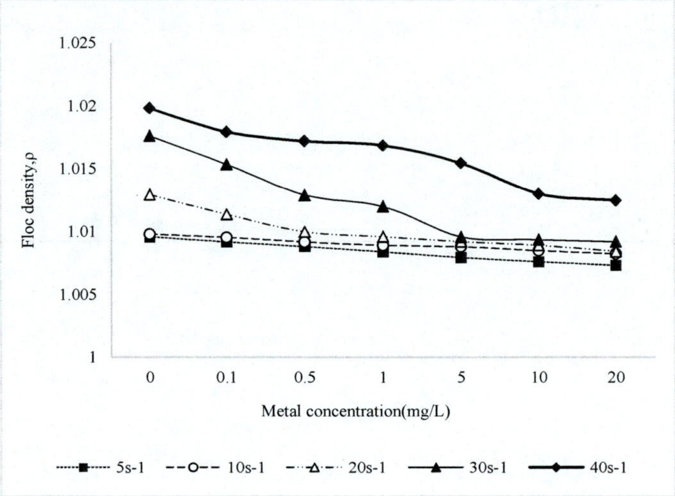

Fig. 6. Variation of floc density with heavy metal concentration for various turbulences

decreases. The larger flocs formed at high metal concentrations possess lower floc density. Macroflocs exhibiting porous structures with low floc density will be easier to get ruptured under turbulence shear. It is observed that the flocs formed at higher turbulence shear (40 s^{-1}) exhibit higher floc density values for all metal concentrations. Higher floc density was also observed for flocs formed at low metal concentrations due to their smaller size compared to those at high metal concentrations. The floc density is negatively related to floc size [6, 18, 20].

3.1.3 Breakage Coefficient

The breakage coefficient indicates the ratio of the number of flocs before the collision to the number of flocs after the collision. In population balance equations a floc is assumed to break into two equal sizes. If the floc breaks into two equal sizes, then its breakage coefficient is taken as ½ which is not true in all cases. The flocs undergo breakage into several sizes yielding different values for breakage coefficient [13]. Figure 7 represents the variation of the breakage coefficient of kaolin with metal concentration at different salinities under different turbulences.

The combined effect of salinity and heavy metal tends to increase the breakage coefficient, making the bond between the macroflocs stronger, and thus making them less liable to disintegration [4] with a minimum breakage coefficient obtained at maximum turbulence of 40 s^{-1}. As the turbulence increases, the macrofloc population decreases due to the breaking up of macroflocs into microflocs, thereby resulting in a decrease in the breakage coefficient. It can be concluded that the maximum breakage coefficient of macroflocs is obtained at higher salinity conditions with a metal concentration of 20mg/l at a turbulence rate of 5 s^{-1}.

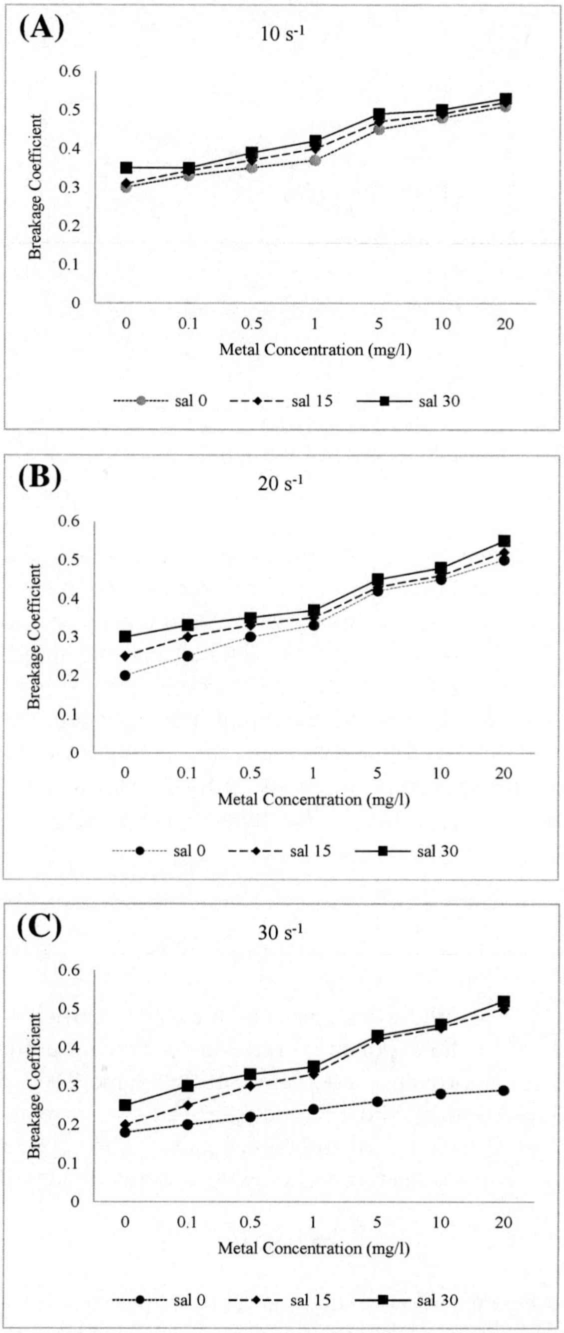

Fig. 7. Variation of breakage coefficient with metal concentration at turbulences (A) 10 s^{-1} (B) 20 s^{-1} (C) 30 s^{-1} (D) 40 s^{-1}.

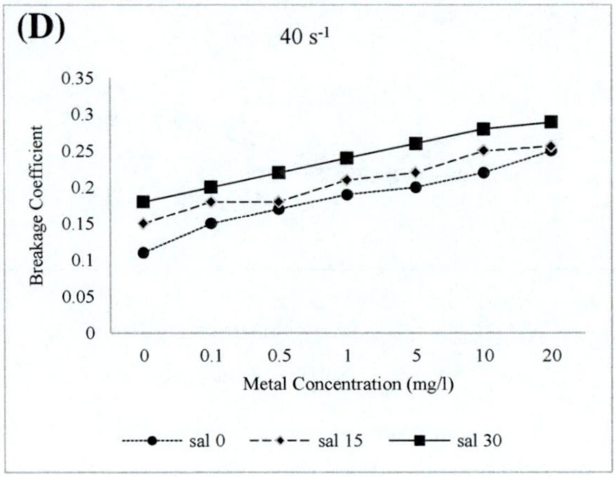

Fig. 7. (*continued*)

In the binary breakage model, the breakage coefficient is found to be 0.5. In Fig. 7, for all salinities and metal concentrations, the breakage coefficient ranges from 0.2 to 0.6. The results suggest that the binary breakage model does not find applicable at turbulences from 5 s^{-1} to 40 s^{-1}. Thus, the detailed analysis of the breakage coefficient from the experimental results led to the development of a relationship for the breakage coefficient in terms of metal concentration for different turbulent rates. Two relationships are developed based on the turbulence range representing lower turbulence and higher turbulence range. An exponential relationship for breakage coefficient Γ in terms of metal concentration M is proposed for turbulence rates ranging from 5 s^{-1} to 10 s^{-1}.

$$\Gamma = 0.270e^{0.092M} \quad (6)$$

$$\Gamma = 0.170e^{0.080M} \quad (7)$$

The relationship (7) was used to estimate the breakage coefficient of flocs due to turbulences above 20 s^{-1}. The exponential relation predicted the breakage coefficient of flocs fairly well with a correlation coefficient of 0.983 and 0.973 respectively, thereby highlighting the effectiveness of the relationship. Another exponential relation for the breakage coefficient Γ in terms of turbulence shear G for clay particles without the presence of heavy metal has been proposed with a correlation coefficient of 0.966.

$$\Gamma = 0.451e^{-0.21G} \quad (8)$$

The breakage coefficient for lower shear rates (10 s^{-1}) ranges from 0.3 to 0.6 while the breakage coefficient for higher shear rates (40 s^{-1}) ranges from 0.1 to 0.3. At average turbulence shear condition, i.e. 20 s^{-1} for all salinity conditions with the highest metal concentration (20 mg/l), the breakage coefficient approximates 0.5 during which the binary coefficient finds application. Thus, the study identified that at a low turbulence level of < 20 s^{-1} and high turbulence level of >20 s^{-1} the binary coefficient model cannot be applied.

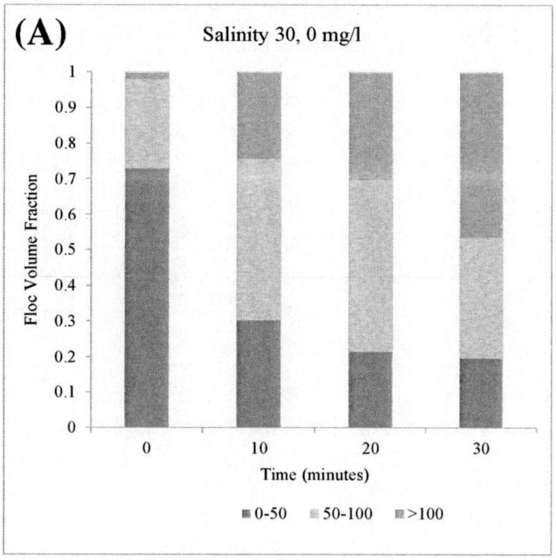

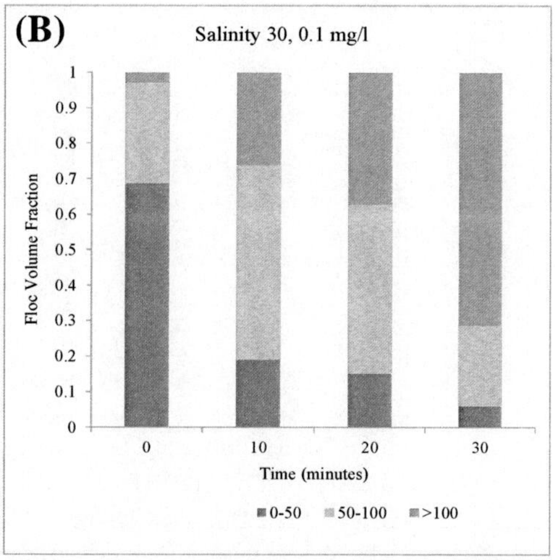

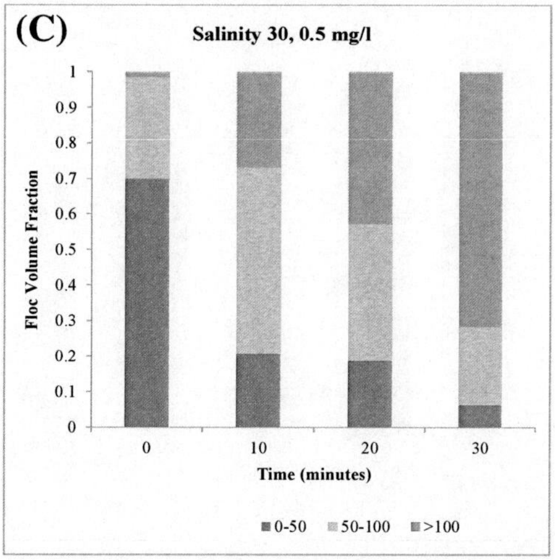

Fig. 8. Variation of floc volume fraction with time interval at metal concentrations (A) 0 mg/l (B) 0.1 mg/l (C) 0.5 mg/l (D) 1 mg/l (E) 5 mg/l (F) 10 mg/l (G) 20 mg/l.

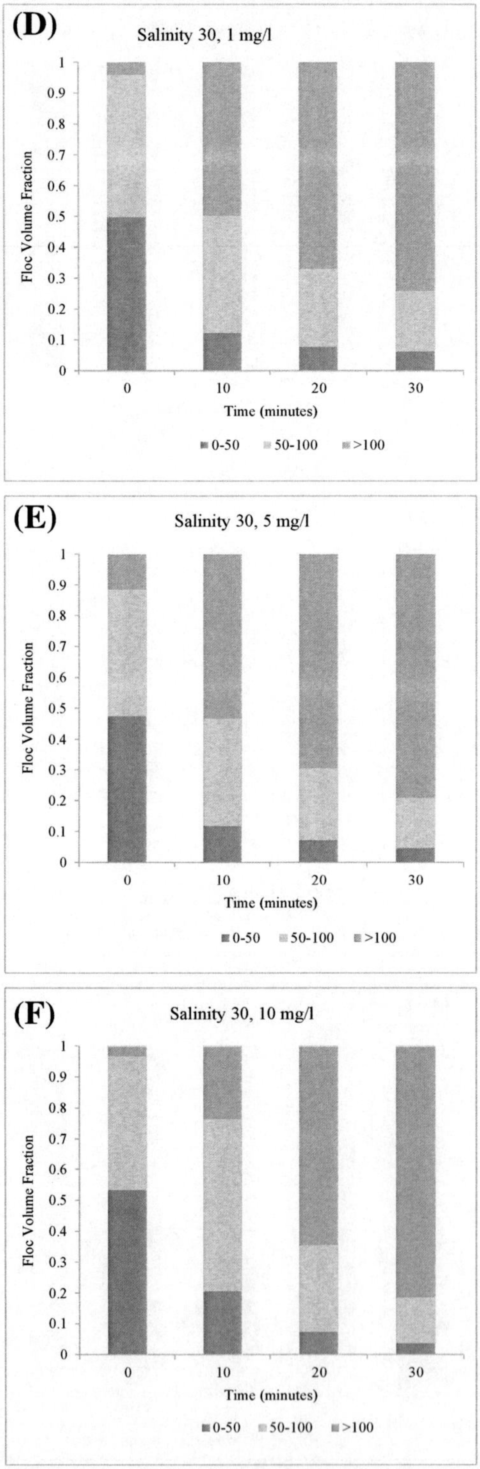

Fig. 8. (*continued*)

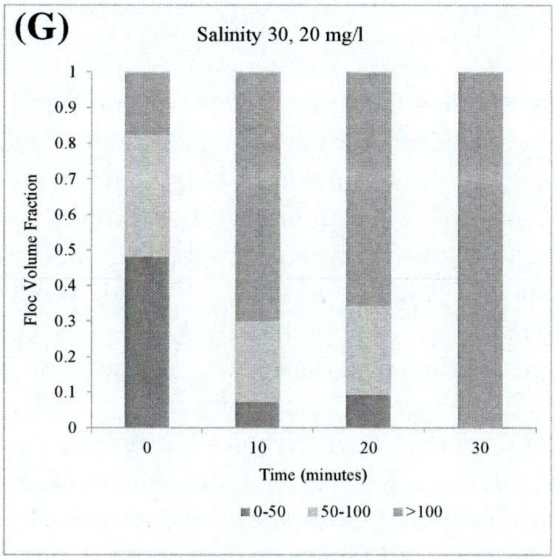

Fig. 8. *(continued)*

3.1.4 Variation of Floc Volume Fraction with a Time Interval for Different Metal Concentration

The variation of floc volume fraction of different classes of flocs with time intervals ranging between 0 to 30 min under the influence of different metal concentrations at salinity 30% with a turbulence rate of 5 s^{-1} is shown in Fig. 8.

On comparing the floc volume fraction of the 3 classes at different salinities, salinity 30 g/l attains to have the maximum volume fraction at the lowest shear rate i.e., 5 s^{-1}. Thus the condition preferred here for obtaining the settling is a turbulence shear rate of 5 s^{-1} at salinity 30. From Fig. 10(G) it is observed that the size class III (>100 μm) increases with an increase in metal concentration and attains the maximum value at 20 mg/l. The size class II (50–100 μm) tends to decrease with an increase in metal concentration. The size class I (0–50 μm) decrease with an increase in metal concentration and reaches the minimum value as shown in Fig. 4(A). As time increases from 0 to 30 min the size class I tends to settle down showing a reduction in their volume fraction.

The decrease of class II particles also indicates the settling of particles from 0 to 30 min. Thus it can be concluded that with the increase in metal concentration and salinity condition, the microparticles settle at a higher rate than the macroparticles.

From the observations, it can be ascertained that with the increase in time floc volume fraction of macro flocs is increasing. The increase in macroflocs is pointing towards hindered settling, wherein the flocs settle as a layer or network. The overlying particles combine with the particles of the same trajectory and settle as a whole [19]. Here, in this case, larger macroflocs have undergone settlement initially and the remaining microflocs are settling as a network.

4 Conclusions

The present study investigates the effects of metal concentration on the flocculation behavior of kaolin under different turbulences and salinity conditions. The presence of heavy metals contributes to the formation of bigger and stronger flocs that are settled and removed easily from the water column. Low turbulent conditions are favorable for floc formation and floc growth, whereas higher turbulences cause the breakage of flocs. It can be ascertained that the presence of metals produces stronger flocs with high breakage coefficients that are comparatively less susceptible to breakage. High values of the breakage coefficient indicate the resistance of flocs to breakage which was observed for high metal concentration and salinity. Finally, a relationship between metal concentration and breakage coefficient for low turbulence shear as well as the high turbulent condition was derived. Further, a relationship between the breakage coefficient and turbulence shear was proposed. Thus, the developed model can be applied to quantify the breakage coefficient under all ranges of turbulence shear encountered in estuaries. Finally, it can be ascertained that higher metal concentration and salinity, hindered settling of particles were observed with an increase in time.

References

1. Argaman, Y., Kaufman, W.J.: Turbulence and flocculation. J. Sanitary Eng., ASCE **96**, 223–232 (1970)
2. Biati, A., Karbassi, A.R.: Comparison of controlling mechanisms of flocculation processes in estuaries. Int. J. Environ. Sci. Technol. **7**(4), 731–736 (2010). https://doi.org/10.1007/BF03326182
3. Bryan, G.W.: The effects of heavy metals (other than mercury) on marine and estuarine organisms. Proc. R. Soc. (B) **177**, 389–410 (1971)
4. Cui, J., Zheng, L., Deng, Y.: Emergency water treatment with ferrate (VI) in response to natural disasters. Environ. Sci.: Water Res. Technol. **4**(3), 359–368 (2018)
5. Eisma, D.: Flocculation and de-flocculation of suspended matter in estuaries. Neth. J. Sea Res. **20**(2/3), 183–199 (1986)
6. Krone, R.B.: A Study of Rheological Properties of Estuarial Sediments. Report No. 63–68, Hyd. Eng. Lab. and Sanitary Eng. Lab. University of California, Berkeley (1963)
7. Manning, A.J., Baugh, J.V., Spearman, J.R., Whitehouse, R.J.S.: Flocculationsettling characteristics of mud: sand mixtures. Ocean Dyn. **60**(2), 237–253 (2010)
8. Maggi, F.: Flocculation dynamics of cohesive sediments. Ph.D. Thesis, TU Delft, 136 p (2005)
9. Mhashhash, A., Bockelmann-Evans, B., Pan, S.: Effect of hydrodynamics factors on sediment flocculation processes in estuaries. J. Soils Sediments **18**(10), 3094–3103 (2017). https://doi.org/10.1007/s11368-017-1837-7
10. Meybeck, M., Helmer, R: The quality of rivers from pristine stage to global pollution. Glob Planet Change **1,** 283–309 (1989)
11. Portela, L.I., Ramos, S., Trigo-Teixeira, A.: Effect of salinity on the settling velocity of fine sediments of a harbor basin. In: Conley, D.C., Masselink, G., Russell, P.E., O'Hare, T.J. (eds.) Proceedings 12th International Coastal Symposium (Plymouth, England), Journal of Coastal Research, Special Issue No. 65, pp. 1188–1193 (1989). ISSN: 0749-0208
12. Priya, K.L., Jegathambal, P., James, E.J.: On the factors affecting the settling velocity of fine suspended sediments in a shallow estuary. J. Oceanogr. **71**(2), 163–175 (2015). https://doi.org/10.1007/s10872-014-0269-x

13. Priya, K.L., Adarsh, S., Venu Chandra, S., Haddout, I., M.S.: Implications of turbulence shear by non-cohesive sediments on the break-up of kaolin flocs. Regional Stud. Marine Sci **39**(2020), 101427 (2020)
14. Ramalingam, S., Chandra, V.: Determination of suspended sediments particle distribution using image capturing method. Mar. Georesour. Geotechnol. **36**(8), 867–874 (2017)
15. Sang, Y, Englezos, P.: Flocculation of precipitated calcium carbonate (PCC) by cationic tapioca starch with different charge densities. I: experimental. Colloids Surf. A Physicochem. Eng. **414**, 512–519 (2012)
16. Sang, Y., Xiao, H.: Clay flocculation improved by cationic poly (vinyl alcohol)/anionic polymer dual-component system. J. Colloid Interf. Sci. **326**(2), 420–425. Engineering Aspects **414**, 512–519
17. Van Leussen, W.: The variability of settling velocities of suspended fine-grained sediment in the Ems estuary. J. Sea Res. **41**, 109–118 (1999)
18. Winterwerp, J.C.: On the dynamics of high-concentrated mud suspension, Ph.D. University of Technology (1999)
19. Winterwerp, J.C.: On the flocculation and settling velocity of estuarinemud. Continent. Shelf Res. **22**, 1339–1360 (2002)
20. Winterwerp, J.C., Manning, A.J., Martens, C., de Mulder, T., Vanlede, J.: A heuristic formula for turbulence-induced flocculation of cohesive sediment. Estuar. Coast. Shelf Sci. **68**, 195–207 (2006)
21. Liu, X., Yin, H., Zhao, J., Guo, Z., Liu, Z., Sang, Y.: Understanding the coagulation mechanism and floc properties induced by Fe(VI) and FeCl3: population balance modeling. Water Sci. Technol. **83**(10), 2377–2388 (2021)

Ecosystem Approach for Sustaining Water Resources

Tri Retnaningsih Soeprobowati[1,2,3](✉), Jumari Jumari[2,3], Riche Hariyati[2,3], and Alam Dilazuardi[3]

[1] School of Postgraduate Studies, Universitas Diponegoro, Semarang, Indonesia
trsoeprobowati@live.undip.ac.id
[2] Cluster for Paleolimnology (CPalim), School of Postgraduate Studies, Universitas Diponegoro, Semarang, Indonesia
[3] Department of Biology, Faculty of Science and Mathematics, Universitas Diponegoro, Semarang, Indonesia

Abstract. Water quality is one of the challenges related to human activities, which in such part, reduces ecosystem services. Ecosystem services are the contribution of ecosystems to human beings. Ecosystem services of most aquatic ecosystems in Indonesia are water provisioning, fisheries, irrigation, recreation, hydroelectricity power, flood protection, erosion prevention, habitat for biodiversity, and socio-culture-religion. An ecosystem has to provide support for living organisms and has to be capable of withstanding pressure. However, when the pressure is too high, it influences the performance of ecosystem services. Sediment samples were collected from 4 sites in Cebong Lake, Dieng, Central Java. Diatoms were separated from sediment with strong acid. Identification was performed using a microscope at 1000 magnification. Based on the diversity index of diatom, the ecosystem of Cebong Lake was relatively stable. Based on the Trophic Diatom Index (TDI), Cebong Lake is in a mesotrophic state (the TDI value is 40–60). Understanding the structure and functioning of lake ecosystems by relating diversity and stability. Regarding water resource management, biodiversity can be applied to evaluate realistic target achievement, ecosystem services, and human impacts. At the ecosystem level, biodiversity expresses natural variability reference conditions.

Keywords: Ecosystem services · Sustainable water management · Biodiversity · Dieng

1 Introduction

In the ecosystem, the interaction between biotic and abiotic components occurs to maintain its stability, called homeostasis. In the ecosystem also occurs the interaction between the species. It was stated that 60% of the ecological systems are globally degraded [1].

Anthropogenic activities have a higher impact on the environment compared to natural activities, which in turn, affected ecosystem services. This approach had been applied in the arid and semi-arid ecological region in China the ecological policy formula

S. Haddout et al. (Eds.): CCORE 2022, SPEES, pp. 102–112, 2023.
https://doi.org/10.1007/978-3-031-26967-7_8

resulting from this research may be adopted by policymakers on solving the global problem of climate change [2].

Human activities alter ecosystem function, particularly in Asia, followed by Europe and America [3], and impacted the species and ecological processes [4]. The ecosystem had 3 functions: stocking energy and material, fluxing energy and material processing, and stabilizing stock rate [5]. However, Groot et al. [6] classified ecosystems into four groups: regulatory, habitat, production, and information functions. Ecosystem attributes to community energetics, community structure, nutrient cycling, catch, and mean trophic level of catch [7].

The Ecosystem services in the water ecosystem provide direct and indirect benefits to humans [8, 9] and can be grouped into four categories: Supporting services—materials necessary for the production of other ecosystem services, such as nutrient cycling; Provisioning services—products gained from ecosystems, such as food and water; Regulating services—benefits obtained from processes such as air or climate regulation, controlling flood and disease; and Cultural services—nonmaterial benefits derived by humans, such as spiritual, recreational, and cultural benefits.

The ecosystem provides a habitat for the organism [10], supporting food chains [11], fisheries production [12], regulating essential ecological processes [13], life support systems and stability [14], nutrient and biogeochemical cycling [15], providing clean air, water, and soil for human [16, 17], regulating climate and soil formation [18].

The ecosystem provides many services that directly and indirectly benefit humans [19]. A key species or individual species has a crucial value in conservation management [20] and can determine ecosystem service's vulnerability [21].

Diatom has been used widely for the assessment of the ecological status of rivers, streams, and lakes [22]. The use of diatoms as biomonitoring tools is well established in many countries [23], but it is a new approach and is currently being used in Indonesia as an indicator for water quality assessment. Diatom used as indicators of the quality of the water environment caused their high sensitivity and preservability and have specific optimal levels and degrees of tolerance to some indicators of the water environment (such as pH, temperature, salinity, and nutrients) [24].

Several studies highlighted that influence of anthropogenic disturbances and natural factors causes the diversity of diatoms to decrease and tends to dominate with several species [25, 26]. Water quality and quantity deterioration are problems in many lakes that require suitable management. This research was done to develop sustainable water management for Cebong Lake based on the ecosystem approach.

2 Methods

2.1 Study Area

Cebong Lake is one of the tourist destinations in Wonosobo Regency, Central Java. Cebong Lake is surrounded by Pakuwojo, Sarojo, and Sikunir Mountain, its water supply is from rain and springs at the bottom of the lake. Since in the early of 1900s, Cebong Lake is used for agriculture (Fig. 2), camping ground and residents who are adjacent to the lake. Previously, Cebong Lake was 12 ha and reduce into 8 ha with the catchment area is 98.58 ha. The lake's average depth is 2.25 m, and the maximum depth is 4.05 m [27].

Benthic diatoms were sampled according to the guidance protocol [28] from 4 research sites (Fig. 1). Site 1 was an area near the potato plantation, Site 2 was the area near the camping ground, Site 3 was the area close to the inlet and potato plantation, and Site 4 was the outlet area. Sediment samples were collected from the upper surface using a shovel and placed in plastic bags. Epipelic diatoms were collected from surface sediments following the standard protocol [29].

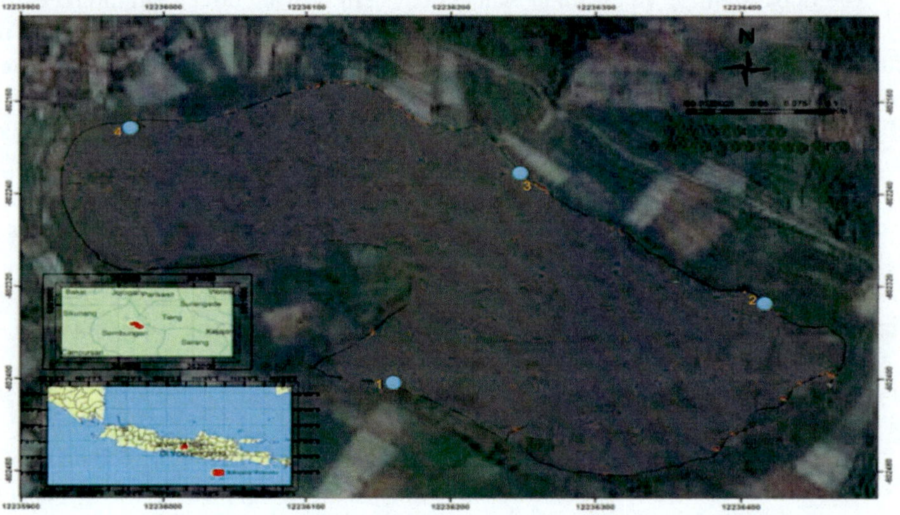

Fig. 1. Four research sites in Cebong Lake, Dieng

Fig. 2. Land use change in the catchment area, and pumped water for irrigation in Cebong Lake

2.2 Diatom Analysis

Epipelic samples were cleaned following the oxidation technique by Soeprobowati et al. [30] with hot acid digestion using 10% HCL and subsequently heated at 100 °C for 2 h with 10% H_2O_2 2 with 90–95 °C in order to remove organic material. The sample was rinsed with aquadest every 6 h and repeated at least 6 times until pH reached neutral, the sample was pipetted at least 400 onto a coverslip and permanently mounted with Naphrax index refractive 1.7.

Diatoms are identified using a microscope Olympus CX23, using identification books [28, 31–36].

2.3 Data Analysis

The diversity index of diatoms, The Shannon-Wiener formula was used [37]:

$$H' = -\sum_{i=1}^{n} pi \ln pi \tag{1}$$

$$p_i : n_i/N \tag{2}$$

where:

H': diversity index

n_i: Number of individuals of type i

N: the total number of individuals of all types

Diversity index criteria:

H' < 1: Low diversity level

1 < H' < 3: Medium diversity level

H' > 3: High lever of diversity

The evenness of the diatom type, it can be calculated using a formula [37]:

$$E = \frac{H'}{Hmax} \tag{3}$$

where:

E:Evenness

Hmax: In s

S: number of species

Evenness index criteria:

E < 0,4 = Low type evenness

0,6 ≥ E ≥ 0,4 = Medium type evenness

E > 0,6 = High type evenness

The dominance of diatom types, was calculated with the formula [37]:

$$C = \sum (P_i)^2 = \left(\frac{n_i}{N}\right)^2 \tag{4}$$

where:

ni: Number of individuals to i

N: The total number of individuals of all types

Dominance index criteria:

D = 0: the absence of predomninant species or stable community structures

D = 1: some species have dominated other species or community structures that change due to ecological pressures

2.4 Trophic Diatom Index (TDI) Analysis

TDI is one of the diatom index for the trophic state of the water. TDI can be calculated based on weighted mean sensitivity (WMS) [38].

$$WMS\frac{\sum_{j=1}^{n} aj.sj.vj}{\sum_{j=1}^{n} aj.vj} \tag{5}$$

where aj = the proportion j valve at the sample
sj = pollution sensitivity
vj = species indicator value

$$\text{TDI} = (\text{WMS} * 25) - 25 \tag{6}$$

Criteria TDI:
0 – 20 = distrophic, very low nutrient content
21 – 40 = oligotrophic, low nutrient content
41 – 60 = mesotrophic, medium nutrient content
61 – 80 = eutrophic, high nutrient content
81 – 100 = hipertrophic, very high nutrient content

3 Results

As a biotic component, the Shannon-Wiener diversity index of diatoms indicated that the ecosystem of Cebong Lake was medium to highly stable (2.74–3.22, Fig. 2). The higher Evenness index was found at Site TC1 (0.68), and the lowest was found at TC2 (Fig. 3). The dominance indices were in the range of 0.07–0.11, indicated no dominant species (Fig. 4).

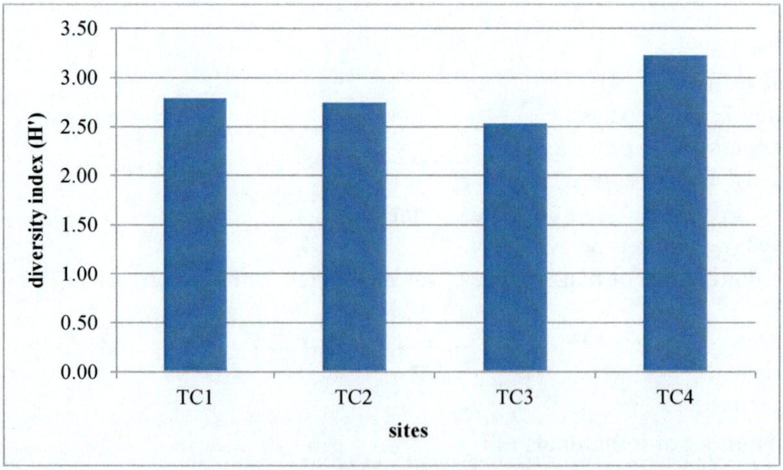

Fig. 3. The Shannon-Wiener diversity index (H') in Cebong Lake, Dieng

Trophic diatom index (TDI) varied from 35 to 47 (Fig. 5), indicated Cebong Lake in the mesotrophic condition (the TDI value in the range 41–60).

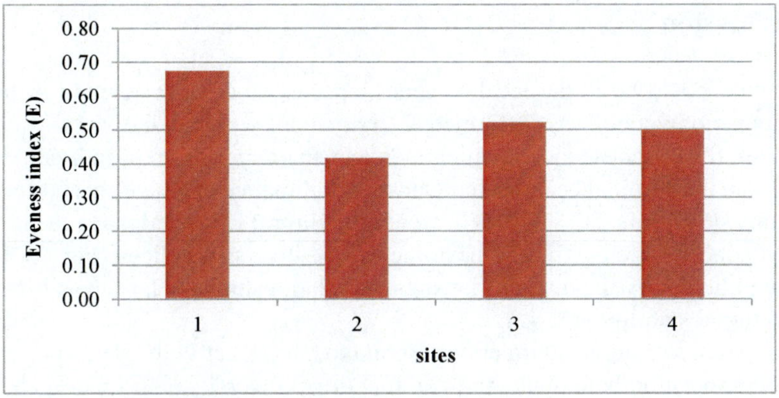

Fig. 4. The Evenness (E) index in Cebong Lake, Dieng

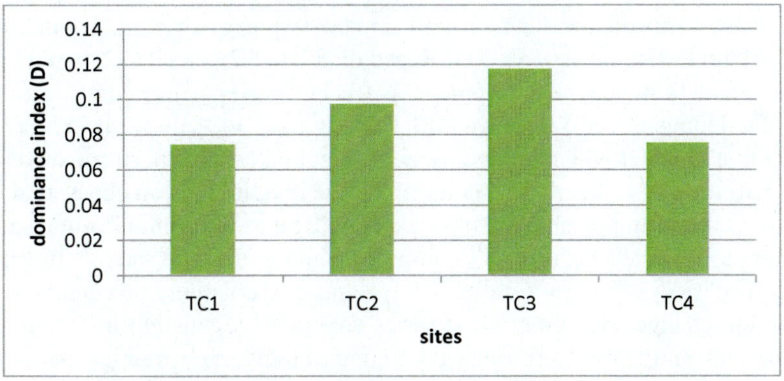

Fig. 5. The dominance index (D) in Cebong Lake, Dieng

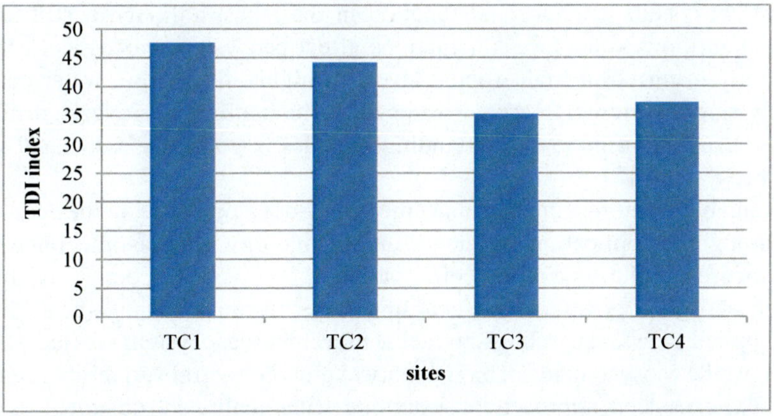

Fig. 6. Trophic diatom index (TDI) from Cebong Lake, Dieng

4 Discussion

Biodiversity could be in the level of genes, species, and ecosystems [39]. In the level ecosystem, biodiversity was performed diversity index which indicates the stability of the ecosystem, the evenness index, which indicates the distribution of individuals between species; and the dominance index indicates the dominance of one of several species in the community [37]. Site TC 1 and TC3 were agricultural impacted, have the highest total nitrogen concentration [40], and have the lowest diversity indices (Fig. 3). Ecological stability indicates by the diversity of stability relationships [41], such as functional and compositional stability [20, 42].

Ecosystem structure and function respond to human activities [19], species richness, species composition/beta diversity [43], functional diversity, and species abundance in the species or community level. Changes in species composition result from change drivers such as species richness, evenness, and dominance [21].

TDI indicated the effect of inorganic nutrients on the river downstream [38]. However, TDI also performed well when applied to lake water quality, such as in Lake Poyang China [44]and Polish lakes [45]. TDI Based on TDI, TC1 and TC 2 were mesotrophic, while TC3 and TC4 were oligotrophic (Fig. 6). TDI well performed of water qualityup-stream of CileungsiRiver, West Java [46], Cebong Lake [40], and Galela Lake [47]. This indicates that the TDI when applied to the tropical lakes, also performed well.

The variation of indices value determined the condition of an ecosystem, therefore, the diversity index and trophic diatom index could be used for monitoring water quality. These indices reflected the impact of anthropogenic to the ecosystem's functioning and services. The ecosystem is not static; it's dynamic and changing constantly. Its structure and function change over time. Time series data provides insight into natural community dynamics and stability through continuous biodiversity monitoring as a base for conservation management [48].

Sustainable water resources may be developed based on this ecosystem approach. The water volume of Cebong Lake was fluctuations. Water level fluctuations were a driver of ecosystem structure and function in the lakes, although it still much to be learned about how water level fluctuations affect ecosystem processes in lakes. Less productivity occurred at high trophic levels in highly fluctuating water compared to relatively stable systems. Seasonal water level fluctuations positively correlated with biomass. Usually, an increase in standing biomass is generally associated with more mature ecosystems [7].

Sustainable water resources management may develop based on the ecosystem approach throughthe application of science and technology, such as integrated water and catchment area, pollution control, conservation of biodiversity; community-based management; and good governance. Regarding water resource management, biodiversity can be applied to evaluate realistic target achievement, ecosystem services, and human impacts. At the ecosystem level, biodiversity expresses natural variability reference conditions. The ecosystem approach for sustaining water quality and quantity was developed for Cebong Lake, Dieng Plateau, Indonesia.

Integrated Water Resources Management (IWRM) combines land and water management through stakeholder participation to realize multiple benefits in watersheds.IWRM is an important climate-adaptation strategy that decreases the vulnerability of freshwater systems to climate change [49].

5 Conclusion

The ecosystem of Cebong Lake was relatively stable based on the diatom diversity index and no domination of one or several species. Based on the TDI Cebong Lake is in a mesotrophic state. Understanding the structure and functioning of lake ecosystems by relating diversity and stability. The effects of direct and indirect human activity on diatom biodiversity and ecosystem functioning. Integrating socio-economic communities into management and governance strategies may develop to promote sustainable use of Cebong Lake.

Acknowledgments. This study was carried out within the framework of Research of a highly reputable international journal (Riset Publikasi Internasional Rereputasi Tinggi, RPIBT) Universitas Diponegoro, contract number 233-38/UN7.6.1/PP/2022**.** Sincerely thank you to Kenanga Sari, Christopher Hardian W, Oki Rachmalia, Wahyu Utari, SafiraRosada, SuyonoLateke, SyarifPrasetyo, and JihanAfifah during fieldwork and laboratory work.

References

1. Huang, C., Huang, X., Peng, C., Zhou, Z., Teng, M., Wang, P.: Land use/cover change in the three gorges reservoir area, China: Reconciling the land use conflicts between development and protection. CATENA **175**, 388–399 (2019). https://doi.org/10.1016/j.catena.2019.01.002
2. Fan, X., et al.: Impacts of human activities on ecosystem service value in arid and semi-arid ecological regions of China. Int. J. Environ. Res. Public Health **18**(21) (2021). https://doi.org/10.3390/ijerph182111121
3. Inácio, M., Barceló, D., Zhao, W., Pereira, P.: Mapping lake ecosystem services: a systematic review. Sci. Total Environ. **847**, 157561 (2022). https://doi.org/10.1016/j.scitotenv.2022.157561
4. Cepic, M., Bechtold, U., Wilfing, H.: Modelling human influences on biodiversity at a global scale–a human ecology perspective. Ecol. Model. **465**, 109854 (2022). https://doi.org/10.1016/j.ecolmodel.2021.109854
5. Kinzig, A.P., Pacala, S.W, Timan, D.: he Functional Consequences of Biodiversity: Empirical Progress and Theoretical Extensions. Princeton University Press. 392p (2002). https://www.jstor.org/stable/j.ctt24hqmf
6. Groot, R.S., Wilson, M., Boman, R.M.J.: A typology for the classification, description and valuation of ecosystem functions, goods and services. Ecol. Eecon. **41,** 393–408 (2002)
7. Gownaris, N.J., Rountos, K.J., Kaufman, L., Kolding, J., Lwiza, K.M.M., Pikitch, E.K.: Water level fluctuations and the ecosystem functioning of lakes. J. Great Lakes Res. **44**(6), 1154–1163 (2018). https://doi.org/10.1016/j.jglr.2018.08.005
8. Costanza, R., d'Arge, R., De Groot, R., Farber, S., Grasso, M., Hannon, B., et al.: The value of the world's ecosystem services and natural capital. Nature **387**, 253–260 (1997)
9. MEA (Millenium Ecosystem Assessment): Overview of the Milliennium Ecosystem Assessment (2005). https://www.millenniumassessment.org/en/About.html
10. Schindler, D.E., Scheuerell, M.D.: Habitat coupling in lake ecosystems. – Oikos 98: 177–189 (2002). https://www.jstor.org/stable/3547424

11. Kwak, I.S., Park, Y.S.: Food Chains and food webs in aquatic ecosystems. Appl. Sci. **10**, 5012 (2020). https://doi.org/10.3390/app10145012
12. Lymer, D., Teillard, F., Carolyn, O.: Land and Water for protein and the micronutrients contribution in the lower Mekong River basin and related countries. In: Global Conference on Inland Fisheries, Freshwater, Fish and the Future, Jan 2015, Rome, Italy. 14p (2015) . https://hal.inrae.fr/hal-02743968
13. Bennett, A.F., Haslem, A., Cheal, D.C. Clarke, M.F., et al.: Ecological processes: a key element in strategies for nature conservation. Ecol. Manag. Restor. **10**(3), 192–199 (2009). https://doi.org/10.1111/j.1442-8903.2009.00489.x
14. Ellery, A.: Supplementing closed ecological life support systems with in-situ resources on the moon. Life **11**, 770 (2021). https://doi.org/10.3390/life11080770
15. Maavara, T., Chen, Q., Van Meter, K., et al. : River dam impacts on biogeochemical cycling. Nat. Rev. Earth Environ. **1,** 103–116 (2020). https://doi.org/10.1038/s43017-019-0019-0
16. Smith, P., Ashmore, M.R., Helaina Black, I.J., et al.: REVIEW: the role of ecosystems and their management in regulating climate, and soil, water and air quality. J. Appl. Ecol. **50**, 812–829 (2012). https://doi.org/10.1111/1365-2664.12016
17. Ashraf, M.A., MohdHanafiah, M.: Sustaining life on earth system through clean air, pure water, and fertile soil. Environ. Sci. Pollut. Res. 26, 13679–13680 (2012). https://doi.org/10.1007/s11356-018-3528-3
18. Schallenberg, M., de Winton, M.D., Verburg, P., Kelly, D.J., Hamill, K.D., Hamilton, D.P.: Ecosystem services of lakes. In: Dymond, J.R. (ed.) Ecosystem Services in New Zealand – Conditions And Trends. Manaaki Whenua Press, Lincoln, New Zealand (2013). http://www.riverlake.co.nz/wp-content/uploads/2017/09/Schallenberg-et-al-2013-ecosystem-services-of-lakes.pdf
19. Costanza, R., De Groot, R., Braat, L., Kubiszewski, I., Fioramonti, L., Sutton, P., et al.: Twenty years of ecosystem services: how far have we come and how far do we still need to go? Ecosyst. Serv. **28**, 1–16 (2017). https://doi.org/10.1016/j.ecoser.2017.09.008
20. White, L., O'Connor, N.E., Yang, Q., Emmerson, M.C., Donohue, I.: Individual species provide multifaceted contributions to the stability of ecosystems. Nat. Ecol. Evol. **4**, 1594–1601 (2020). https://doi.org/10.1038/s41559-020-01315-w
21. Ross, S.R.P.J., Arnoldi, L.M., White, C.D., Stout, J.C., Jackson, A.L., Donohue, I.: Universal scaling of robustness of ecosystem services to species loss. Nat. Commun. **12**, 5167 (2021). https://doi.org/10.1038/s41467-021-25507-5
22. Blanco, S., Cejudo-Figueiras, C., Álvarez-Blanco, I., van Donk, E., Gross, E.M, Hansson, L.A., et al.: Epiphytic diatoms along environmental gradients in Western European shallow lakes. Clean—Soil, Air, Wate*r* **42,** 229–235 (2014). https://doi.org/10.1002/clen.201200630
23. Harding, W.R., Taylor, J.C.: Diatoms as indicators of historical water quality: a comparison of samples taken in the Wemmershoek catchment (Western Province, South Africa) in 1960 and 2008. Water SA **40**(4), 601–607 (2014). https://doi.org/10.4314/wsa.v40i4.4
24. Liu, S., Yao, M., Chen, S., Yuan, X.: Surface sediment diatom assemblages response to water environment in Dongping Lake. North China. Water. **13**(339), 1–13 (2021). https://doi.org/10.3390/w13030339
25. Feret, L., Bouchez, A., Rimet, F.: Benthic diatom communities in high altitude lakes: a large-scale study in the French Alps. Ann. Limnol- Int. J. Lim. **53**(1), 411–423 (2017). https://doi.org/10.1051/limn/2017025
26. Boggero, A., Zaupa, S., Musazzi, S., Rogora, M., Dumnicka, E., Lami, A.: Environmental factors as drivers for macroinvertebrate and diatom diversity in Alpine lakes: New insights from the Stelvio National Park (Italy). J. Limnol. **78** (1019). https://doi.org/10.4081/jlimnol.2019.186

27. Sudarmadji, Supriyanto, H., Lestari, S.: Volcanic lakes in Dieng Plateau: their uses and problems Teknosains. **5** (1), 36 (2015). https://doi.org/10.22146/teknosains.26856
28. Taylor, J.C., Harding, W.R., Archibald, C.G.M.: An Illustrated Guide to Some Common Diatom Species from South Africa. Water Research Commission, South Africa (2007)
29. CEN: Water quality - Guidance standard for the routine sampling and pretreatment of benthic diatoms from rivers. EN. In: European Committee for Standardization. Brussels. 13946 (2005). https://standards.iteh.ai/catalog/standards/cen/8e62f4b7-732a-4197-9efa-681b339b3740/en-13946-2003
30. Soeprobowati, T.R., Tandjung, S.D., Sutikno, Hadisusanto, S, Gell. P.: The minimum number of valves for diatoms identification in Rawapening Lake, Central Java. BIOTROPIA. **23**(2), 96–104 (2016). http://orcid.org/0000-0001-7525-7028
31. Kramer, K., Lange-Bertalot, H.: Susswasserflora Von Mitteleuropa, Bd. 02/1: Bacillariophyceae: Teil 1: Naviculae. Spectrum, Berlin (2004)
32. Kramer, K., Lange-Bertalot, H.: Susswasserflora Von Mitteleuropa, Bd. 02/2: Bacillariophyceae: Teil 2: Bacillariphyceae, Epithemiaceae, Surirellaceae. Spectrum, Berlin (2004)
33. Kramer, K., Lange-Bertalot, H.: Susswasserflora Von Mitteleuropa, Bd. 02/3: Bacillariophyceae: Teil 3: Centrales, Fragillariaceae, Eunotiaceae. Spectrum, Berlin (2004)
34. Gell, P.A., Sonneman, J.A., Illman, M.A., Sincock, J.A.: An Illustrated Key to Common Diatom Genera from Southern Australis. Cooperative Research Centre for Fresh Waater Ecology Identification Guide No. 26, Albury (1999)
35. Guiry, M.D., Guiry, G.M.: AlgaeBase. World-wide electronic publication, National University of Ireland, Galway. https://www.algaebase.org. Accessed 2 Jan 2022
36. Diatom of North America. diatoms.org. .access 12 September 2022
37. Magurran, A.E.: Measuring Biological Diversity. Blackwell, Oxford, 256pp (2004). http://www.bio-nica.info/Biblioteca/Magurran2004MeasuringBiological.pdf
38. Kelly, M.G., et al.: The Trophic Diatom Index: A User's Manual. Environment Agency, Rio House, Waterside Drive, Aztec West, Almondsbury, BRISTOL, BS32 4UD (2001). http://ea-lit.freshwaterlife.org/archive/ealit:4414/OBJ/10375_ca_object_representations_media_197_original.pdf
39. Wilson, E.O.: Biodiversity. Washington, DC: The National Academies Press. 538p (1997). https://doi.org/10.17226/989
40. Soeprobowati, T.R., Purnaweni, H., Jumari, J., Sari, K.: The relationship of water quality to Epipelic Diatom Assemblages in Cebong Lake, Dieng Indonesia. Polish J. Environ. Stud. **31**(1), 281–295 (2022). https://doi.org/10.15244/pjoes/137084
41. Wang, S., Loreau, M., Mazancourt, C., de Isbell, F., Beierkuhnlein, C., Connolly, J., et al.: 602 Biotic homogenization destabilizes ecosystem functioning by decreasing spatial asynchrony. Ecology **102,** e03332 (2021). https://doi.org/10.1002/ecy.3332
42. Hillebrand, H., Kunze, C.: Meta-analysis on pulse disturbances reveals differences in functional and compositional recovery across ecosystems. Ecol. Lett. **23**(3), 575–585 (2022). https://doi.org/10.1111/ele.13457
43. Magurran, A.E., et al.: Divergent biodiversity change within ecosystems. Proc. Natl. Acad. Sci. **115**, 1843–1847 (2018). https://doi.org/10.1073/pnas.1712594115
44. Yang, J., et al.: Applicability of benthic diatom indices combined with water quality valuation for Dish Lake from Nanjishan Nature Reserve. Lake Poyang. Water **12**(10), 2732 (2020). https://doi.org/10.3390/w12102732
45. Gebler, D., Kolada, A., Pasztaleniec, A., Szoszkiewicz, K.: Modelling of ecological status of Polish lakes using deep learning techniques. Environ. Sci. Pollut. Res. **28**(5), 5383–5397 (2020). https://doi.org/10.1007/s11356-020-10731-1
46. Pasisingi, N., Pratiwi, N.T.M., Krisanti, M.: The use of trophic diatom index To...in The Upstream of Cileungsi River, West Java. Ind. Fish. Res. J. **20**(1), 11 (2014). https://doi.org/10.15578/ifrj.20.1.2014.11-16

47. Soeprobowati, T.R., Saraswati, T.R., Jumari, J. et al.: Diatom index of Galela Lake, Halmahera, Indonesia in relation to human activities. Int. J. Environ. Sci. Technol. (2022). https://doi.org/10.1007/s13762-022-04463-7
48. Keitt, T.H., Abelson, E.S.: Ecology in the age of automation. Science **373**, 858–859 (2021). https://doi.org/10.1126/science.abi4692
49. Kundzewicz, Z.W., Mata, L.J., Arnell, N.W., Döll, P., et al.: The implications of projected climate change for freshwater resources and their management. Hydrol. Sci. J. **53**(1), 3 (2008). https://doi.org/10.1623/hysj.53.1.3

Flood Modelling and Inundation Mapping of Meenachil River Using HEC-RAS and HEC-HMS Software

S. Athira(✉), Yashwant B. Katpatal, and Digambar S. Londhe

Department of Civil Engineering, Visvesvaraya National Institute of Technology, Nagpur, Maharashtra, India
athirajuly31@gmail.com, digambarlondhe@students.vnit.ac.in

Abstract. The flood modelling and inundation mapping study was done for the Meenachil river basin in Kerala. The study shows the usage of HEC-RAS and HEC-HMS tools in determining the flood hazard and vulnerability. Flood inundation maps for different return periods were developed to analyze the future flooding scenarios. Abnormal heavy rainfall in July-August 2018 which leads to severe flooding caused loss of many lives and damages to properties in Kerala. This study also analyzed the 2018 flood event to develop the flood inundation maps and to identify the flood affected area of different land use classes as well as the flood affected road infrastructure. The Meenachil river basin was delineated using HEC-HMS and the topographic characteristics were extracted from SRTM DEM. Hydrologic model was developed using SCS Curve number method, SCS unit hydrograph method and Muskingum method for loss, transform and routing method respectively. The simulated hydrologic model was manually calibrated using discharge data. The precipitation data from the year of 1990 to 2017 was used to run the hydrologic simulation to obtain the discharge values which were used as input for hydraulic analysis using HEC-RAS. The discharge values obtained were then used to find the peak discharge for different return periods of 5, 10, 50, 100, 500 years using Gumbel's method. The simulated discharge is fed to HEC-RAS as the upper boundary condition for identifying the flood affected area. One dimensional steady state analysis was carried out to obtain the flood depth for different return periods. Similarly, hydraulic analysis was carried out for the year of 2018. The peak annual discharge in cumecs was calculated using Gumbel's method and found to be 7484.468 m^3/s. The flood plain map covered an area of 24.2465 km^2. It was observed that depth of flow increased towards the downstream end. The developed flood inundation map was superimposed on road map to find the length of different classes of roads that got submerged. Land use map was generated for the year of 2018 through Isocluster method of unsupervised classification of LISS III image. This was used to identify the area of different land use classes that were affected during 2018 flood as per the simulation.

Keywords: SRTM DEM · HEC-RAS · HEC-HMS · Gumbel's method · Flood inundation mapping

S. Haddout et al. (Eds.): CCORE 2022, SPEES, pp. 113–130, 2023.
https://doi.org/10.1007/978-3-031-26967-7_9

1 Introduction

Over the past few years, it has been observed that around 300 natural disasters occur globally affecting millions of people. Flood is one of the most catastrophic and frequent natural disaster that claims thousands of lives and causes lot of damages to the properties, interrupting communications and transportation all over the world. Based on the different types of urban regions, flood can be divided into riverine floods, urban drainage floods, flooding caused by lake levels and coastal flooding. The different geographical makeup and unique topography of an urban area strongly influences the possibility of certain type of flood [1]. Flood risk is related to two elements i.e., flood hazard and flood vulnerability. Between 1998 and 2017, floods devastated about 2 billion people around the world [2]. Studies have shown that rapidurbanization and industrialization combined with the climate change fuelled by the greenhouse gases accounts for a strong tendency towards the extreme precipitation [3, 4].

The effects of flood can be very devastating to a community. Flooding can result in the loss of lives, damage to buildings and infrastructure, and the devastation of crops and livestock in the short term. Interruptions to communication networks and vital infrastructure (such as power plants, roads, and hospitals) are examples of long-term effects that can have severe social and economic consequences.More than 75% of death in flood is due to drowning [2]. Also, large magnitude floods occur at an interval of several years to decades and those floods are competent to change the channel morphology in a significant way [5].

Floods are the most frequent and common catastrophe in India. Flood-prone areas cover more than 40 million hectares (mha) of the overall geographical area of 329 million hectares (mha). In the last ten years, from 1996 to 2005, the average annual flood damage was Rs. 4745 crores, compared to Rs. 1805 crore in the prior 53 years. Floods affect 75 lakh hectares of land on an annual basis, killing 1600 people and causing Rs.1805 crores in damage to crops, residences, and public utilities. Major floods occur every five years or more frequently. Over the last decade, 2018 Kerala floods is a globally reported disaster which affected all 14 districts of the state [6]. Assam witnessed heavy flooding in the Brahmaputra River during the monsoon of 2020. Over five million people were affected by the floods as of October 2020, with 123 people killed and another 26 killed due to landslides. 2016 Brahmaputra floods affected more than 1.8 million people in India as well as fauna of Kaziranga National Park and PobitoraWildlife Sanctuary [7]. Likewise, every year some region of the country is badly affected by floods.

The floods occurred in August 2018 in Kerala were a catastrophe of immense proportions. It is the largest storm to hit the state, according to scientists. The calamity, which occurred in the middle of August, altered the environment on both land and sea, affecting bio resources, livelihood, and capital assets. While 483 people died and countless of animals perished, the disaster caused massive damage to the state's infrastructure, with hundreds of roads, bridges, and homes getting washed away. During the 2018 monsoon, the state was hit by nature's fury in the form of floods and landslides. Kerala experienced 2346.6 mm of rain between June 1 and August 19, according to IMD statistics, compared to a predicted 1649.5 mm. This year's rainfall reached 42% more than usual. Furthermore, rainfall in Kerala was 15%, 18%, and 16% above normal during June, July, and the 1st to 19th of August, respectively. The disaster left a path of devastation across the state with landslides in the hills and floods in the lowlands and plains. The continuous rain from August 8th to 18th was unusually heavy and prolonged. Eleven of the state's 14 districts were seriously impacted, with the Kerala government estimating a loss of Rs 40000 crore (not including the loss of 483 lives and 55,439 hectares of agricultural land) [6]. With significant increase in risk over the past few years, there is an urgent need for collaborative, long-term approaches to mitigate the future impacts. The natural disasters cannot be eradicated but somehow has to be managed efficiently. Management of flood risk is therefore important to reduce the impact of flood events and predict the future flood hazard.

Flood modelling has advanced dramatically in present years and is now an integral part of all systems to forecast flood. Data of large quantity, both temporal and spatial, is necessary for hydrological modelling. Collecting highly precise data into three categories—hydrologic, hydraulic, and topographic—is the first stage in the creation of a foodplain study [8]. Precipitation, stream profiles, river flows, and features of watershed are examples of spatial and temporal geographical data that can be included using GIS. As a result, effective flood modelling requires the combination of hydrological models with geographic information systems (GIS) [1] Some of the popular flood forecasting models are Artificial Neural Networks (ANNs), Hydrologic Models, Hydraulic models and coupled models. Dang Thanh and Florimond combined the Hydrological and Hydraulic Model for Flood Prediction in Vietnam [9]. Thakur et al. Coupled HECRAS and HEC-HMS to evaluate foodplain inundation map in Illinois State in USA [10].

Hydrological modelling is commonly used in evaluating the basin‘s hydrological response to rainfall. It represents the relation between rainfall and runoff to the catchment area. It plays a vital role in water resources planning, management, flood forecasting and in many other applications. The planning, design, construction and operation of hydraulic projects requires an appropriate information about the variation of catchment runoff using various models so that among the different alternative, the one that best fits can be used for the prediction of future responses of the catchments. Depending on the objective of the study and suitability there are several rainfall-runoff models.

HEC-HMS or Hydrological Engineering Centre Hydrological Modelling System is a Hydrological model developed by U.S army corps of Engineers. Recently, GIS has become an integral part of the hydrologic studies [11]. HEC-HMS is a tool that simulates the entire hydrologic process of a dendritic watershed. Many standard hydrologic analysis processes, including event infiltration, unit hydrographs, and hydrologic routing are included in the ArcGIS software. Evapotranspiration, melting of snow and moisture content of soil accounting are among the techniques included in HEC-HMS for continuous simulation. Advanced capabilities for gridded simulation of runoff employing the linear quasi-distributed runoff transform are also available (ModClark). Optimization of model, streamflow predictions, depth-area reduction, analysing uncertainty of model, erosion and transport of sediments and characteristics of water are all provided by additional analysis tools. Narayan et al. compared the peak food value from two food frequency analysis techniques—namely, from Log Pearson type-III and Gumbel's extreme value distribution functions—with the food frequency results obtained from the HEC-HMS model, and they discovered that the model output is marginally higher than the result of the extreme value distribution function [12]. Finally, it was recommended that the

HEC-HMS model's output was more trustworthy than the output from the two food frequency methodologies when compared to the research area's actual conditions. The major demerit of using HEC-HMS is that in order to retrieve the parameters, the model calibration must be updated either automatically or manually. Manual adjustments take up a lot of time.

Hydraulics defines the engineering analysis of the flow of water over natural streams and manmade channels. Hydraulic model is a mathematical model used to analyse the hydraulic behaviour of a liquid flow system such as water systems, sewer flow systems and storm systems [13]. Through hydraulic models, we create a physical representation of the real world using the collected observed dataset. Hydraulic modelling helps in identifying methods to improve the system reliability and efficiency by evaluating the existing or future systems. Hydraulic models can be applied in flood risk analysis and can thereby predict the extent of flood and river water levels for implementing proper flood management measures.

There are a bunch of software packages available that can perform hydraulic computations along the rivers. One of the most accepted and ubiquitous runoff model in the water resources engineering industry is HEC-RAS. The U.S Army Corps of Engineers' River Analysis System [14] developed by Hydrologic Engineering Centre is an integrated software that helps in performing one dimensional steady and unsteady, gradually varied flow river hydraulics calculations, transport of sediments, mobile bed evaluation, water quality analysis, water temperature modelling and several hydraulic design computations [14]. The adoption of a single geometric data structure and shared geometric and hydraulic computing methods for all four components is a crucial feature of the software.

The main objectives of the study are to develop a hydrologic model of the study area and to obtain the required discharge values using HEC-HMS software. Hydraulic model was developed using the obtained discharge and river geometry. Developed model was then used to generate the flood inundation maps for various return periods. Flood inundation map for 2018 flood was studied and the area of different land use classes and length of different road classes submerged during the flood were identified.

2 Methodology

2.1 Study Area

The Meenachil River, the lifeline of Central Kerala, is formed by the confluence of four major streams fed by several lesser tributaries that originate in the Cardamom Hills of the Southern Western Ghats. The perennial river flows through 57 panchayats and three municipal towns in Kottayam as well as the Taluks of Vaikom, Kanjirappally, Meenachil, and Changanacherry, and has a basin size of 1272 km^2. River is about 78 km long. The watershed stretches from 9°25′ to 9°55′ latitude and 76°20′ to 76°55′ longitude. Meenachil is one of the state's undammed rivers, and hence the river's flooding was caused solely by unusually heavy rains in the catchment area, rather than a sudden discharge from reservoirs. The study area of Meenachil River basin is shown in Fig. 1. River basin has good rainfall and humid atmosphere throughout the year due to both south-west and north-east monsoon. It receives an annual rainfall in the range of 2420 mm to 4686 mm. The monthly mean temperature varies from 26.2 °C to 29.4 °C and the wind speed varies from 6.55 to 10.55 km/h.

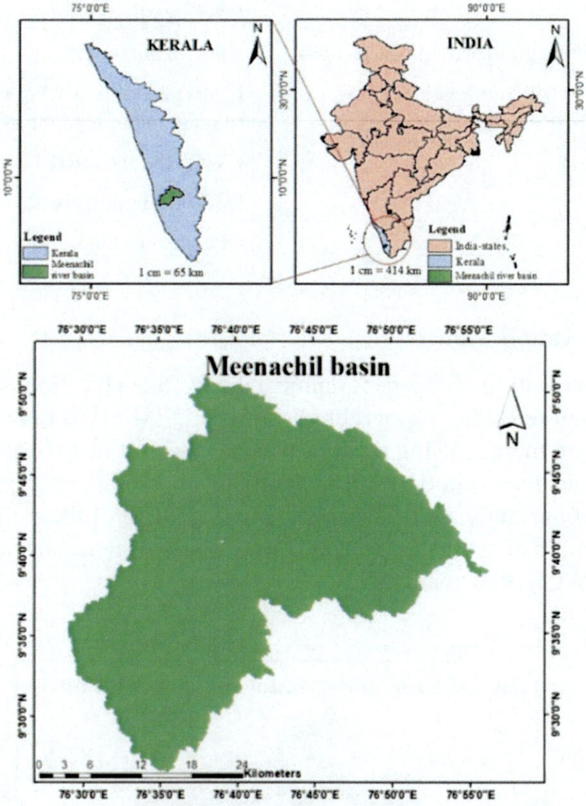

Fig. 1. Study area Meenachil basin

2.2 Data Collected

In any rainfall-runoff modelling, input parameters such as physiographic (Digital elevation models, land use land cover maps and soil maps) and hydro-meteorological (rainfall data and stream flow/ discharge data) databases play a crucial role. The information of the data collected for the study and their sources are given in Table 1.

2.3 Hydro-meteorological Data

In the present study, discharge data in cumecs collected at Kidangoor station from the year 1990 to 2018 was used. This data was used for calibrating the model. The precipitation data for the study area was obtained from CWC for the year of 1990 to 2018. To obtain the discharge value for different return periods, Gumbel method is used.

Table 1. Data collected for the study

Data collected	Source
SRTM DEM	USGS earth explorer
Gridded SCS Curve number map (GCN 250)	www.figshare.com
Daily Discharge and Water levels	Kuniyil, Centre for Water Resources and Management, Calicut. (CWRDM)
Daily Precipitation	CWC (INDIAWRIS)
LISS III Image	ISRO Bhuvan website
Street map	Open Street map, India

2.4 Digital Elevation Model

The primary information on the geographical details like elevation, slope was extracted from SRTM (Shuttle Radar Topography Mission) DEM of 30 m resolution. The individual datasets are merged using the mosaic raster dataset in ArcGIS. The no data cells or voids in the DEM was filled by interpolating in automatic way using raster functions. The boundary of the study area was used to clip the DEM of the study area as shown in Fig. 2. The elevation of watershed ranges from −17 to 1188 m. The projected coordinate system used is WGS1984 zone 43N.

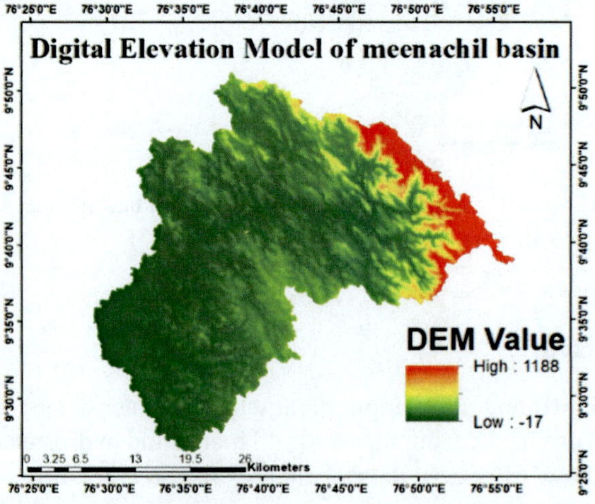

Fig. 2. DEM of the study area

2.5 Global Curve Number Gridded Map

GCN 250 is a gridded dataset defining curve numbers (CN) at the 250 m spatial resolution developed from new global land cover of spatial resolution of 300 m and soils data of 250 m resolution. The grid map was developed in the year of 2018 as shown in Fig. 3.

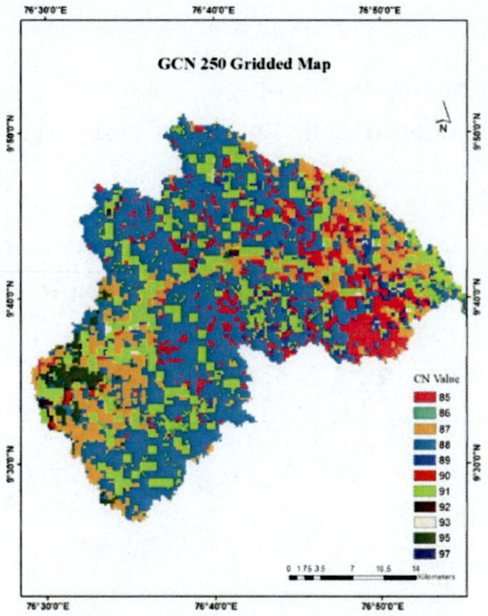

Fig. 3. Curve number gridded map

2.6 LISS III Image

ISRO's LISS-III camera delivers multispectral data using 4 bands. With a ground swath of 141 km, the spatial resolution for visible (two bands) and near infrared (one band) is 23.5 m. LISS III images for the year 2018 were downloaded from the Bhuvan store. The satellite image underwent the pre-processing such as merging and mosaic of multiple tiles in ArcGIS software. The LISS III image was then clipped for the extent area of Meenachil river basin. The Fig. 4 shows the LISS III image of Meenachil River basin in False Colour Composite.

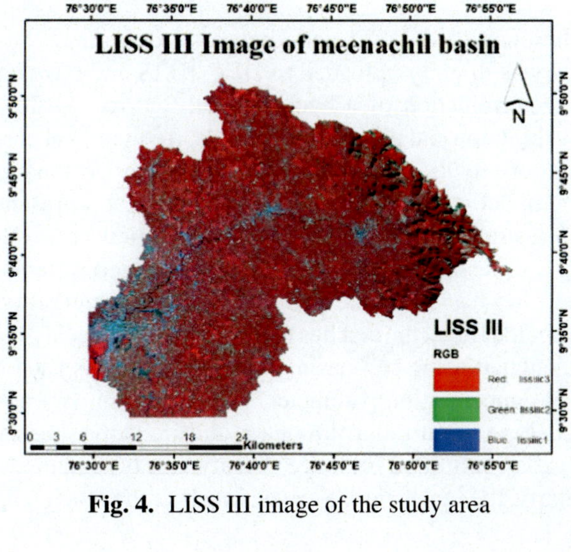

Fig. 4. LISS III image of the study area

2.7 Street Maps

The street map with different classes of roads were downloaded as a shape file from openstreet.org and was clipped to the extent of study area, Meenachil river basin as shown in Fig. 5.

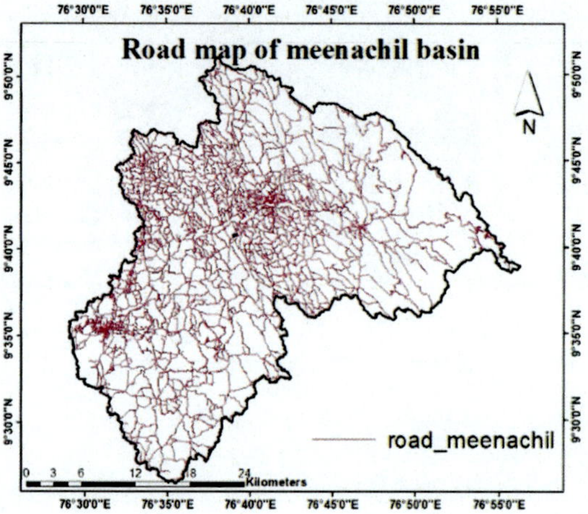

Fig. 5. Street map for the study area

2.8 Methodology Adopted

The following flowchart (Fig. 6) shows the various steps involved in the methodology.

2.9 Hydrologic Modelling Using HEC HMS Software

30" × 30" resolution SRTM DEM was downloaded from U. S. Geological Survey (USGS). The data was directly uploaded to HEC-HMS and terrain pre-processing was carried out. The determination of a hydrologically correct DEM and its derivatives, primarily the flow direction and accumulation grids, frequently necessitates considerable iteration of calculations for drainage paths to accurately reflect the water flow through the basin. The terrain model is used to prepare grid layers that represent the flow direction, flow accumulation, stream network, stream segmentation and watershed delineation, vector layers of the watersheds and streams and aggregated watersheds.

Different small sub-basins are merged using the basin merge tool. River length and River slope is computed for all the reaches using DEM and river layer as input parameters. Basin slope is calculated using sub-basin and slope grid. It is used for the computation of CN and time of concentration parameter. The sub-basin parameters such as longest flow path and its slope, Centroidal flow path and its slope, basin slope, basin relief, relief ratio, elongation ratio and drainage density can be obtained directly after basin processing. Similarly, the reach characteristics such as the reach length and slope, relief

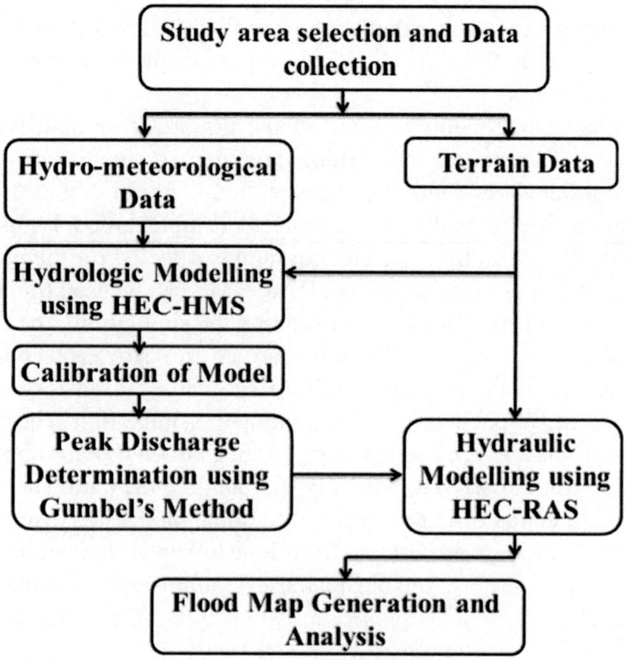

Fig. 6. Flowchart of Methodology adopted in the study

value, sinuosity can be derived as well. These basin characteristics can be used to estimate the hydrologic parameters such as lag time and time of concentration etc.

The hydrologic parameters for HMS processes were computed by providing the Loss method, Transform method and Routing method. In the present study, the SCS curve number (CN) method was adopted for loss method, SCS unit hydrograph (UH) method for transform method and Muskingum method for routing of reaches. The main parameter for SCS CN method is the curve number which was obtained from the global curve number grid map. The CN value for each of the sub basin was determined using zonal statistics tool in ArcMap. The current study used the SCS-UH approach since the only parameter that is not known is the lag time. This model uses a single peaked, dimensionless UH. Calibrating the model with measured data after simulation can be used to predict the time for each of the sub basin. Estimation of lag time can be done using empirical equations such as,

$$t_{lag} = 0.6 \times t_c \tag{1}$$

where, t_c is the time of concentration or the amount of time it takes for water to flow from the farthest point in a watershed to the outflow. The value of the same in hours can be determined from an empirical equation that depends on longest flow length in feet (l), average watershed slope in % (Y) and maximum potential retention in inches (S).

$$t_c = \frac{l^{0.8}(s+1)^{0.7}}{1140Y^{0.5}} \tag{2}$$

The river crossing process was calculated using the Muskingum method. The National Resources Conservation System (NRCS) incorporates the basic parameters x and K in the Muskingum equation.

Meteorological components are one of the important components in the project. The main purpose is to prepare meteorological boundary conditions for the sub basins. Meteorological model is created after the creation of basin model. It includes precipitation, Evapotranspiration and Snowmelt. Meteorological model must include rainfall information for sub basins. Specified hyetograph is selected for the present study.

Control specification manager tool in HEC-HMS can be used for creating as well as for copying the control specifications for running the simulation. The time window over which the simulations are supposed to be performed are set by each control specification. Control specification manager tool in HEC-HMS can be used for creating as well as for copying the control specifications for running the simulation. The time window over which the simulations are supposed to be performed are set by each control specification.

Optimization was carried out manually by changing the parameter values after multiplying them with factors and the performance of the model was evaluated. The precipitation values for four monsoon months, from June to September was used for calibrating the model. The discharge observed during these 4 months were obtained from CWRDM, Kozhikodu, Kerala.

2.10 Hydraulic Modelling Using HEC-RAS

The preliminary step I hydraulic modelling is pre-processing of terrain which can be carried out by RAS-mapper, a GIS tool in new version of HEC-RAS. Pre-processing extracts the physical characteristics of the study reach. The SRTM DEM downloaded from USGS was projected to WGS 1984 UTM zone 43N and clipped to get the required study area.

River centre line, bank line, flow path and cross-sectional cutlines were drawn and the river was digitized. After digitizing the terrain in RAS-Mapper, Geometry data and flow data are the main input parameters needed for the hydraulic simulation. The geometric data of the terrain was edited to input parameter such as Manning's n value. (HEC-RAS manual).

The peak discharge values for different return periods were calculated using Gumbel's method. The Gumbel frequency analysis approach is based on extreme value distribution and employs frequency factors created for theoretical distributionThis approach, developed by Gumbel, is a well-known probability distribution function that is frequently used in hydrological studies to anticipate extreme hydrological events, particularly the maximum projected rainfall and flood peak. The Gumbel technique uses the following procedure to compute the expected maximum discharge (X_T) of the multiple return periods (T) and the probability of exceedance (P).

$$X_T = \overline{x} + K_T \times \delta \tag{3}$$

where, X_T = The maximum value of expected rainfall, $\overline{x}$ = mean rainfall, δ = standard deviation and K_T = frequency factor.

Here, the frequency factor can be calculated from parameters like reduced variate (Y_T), mean of reduced variate ($\overline{Y}$) and standard deviation of reduced variate (δ),

$$K_T = Y_T - \frac{y}{\delta} \tag{4}$$

$$Y_T = -\ln\left(ln\left(\frac{T}{T-1}\right)\right) \tag{5}$$

The following assumptions are made in the software for running the analysis.

- The flow is steady and gradually varied except at hydraulic structures
- Velocity component in the direction of flow is only taken into account i.e., the flow is considered to be 1 dimensional.
- River channel slopes are assumed to be small, less than 1:10. For performing modelling of river channel with steeper slopes, the computed depth of water is divided by cos θ to get the correct depth of water. However, HEC-RAS doesn't take into account phenomenon such as air entrainment in the flow with the increase in slope which could be a limitation.

2.11 Flood Studies

The model developed in HEC-RAS was used to obtain the flood inundation map for the year of 2018 flash flood using the peak discharge obtained. The final stage involves exporting the results of the simulation from RAS-Mapper to Arc-GIS. They are converted from raster format to a shape file by using reclassify tool in Arc-GIS. Then the area of build-up land and cultivable land flooded during 2018 flood are calculated. This can be done by overlaying the inundation map over the land use land cover map developed for the study area. LISS III image for the study area was downloaded from ISRO Bhuvan store and the different tiles were mosaicked after composing the 4 bands. LULC map is then generated using Iso cluster unsupervised classification method. Iso cluster method can recognise clusters within an image and then classify them as a single class. The HEC-RAS output shape file is then used as an extent to clip the LULC raster. The area of the resultant classes in the clipped LULC is calculated using the zonal statistics tool in Arc-GIS. The street map from the year of 2018 obtained from Open Street Map was used to identify the length of different classes of streets that got flooded in the year of 2018 using the similar procedure.

3 Results and Discussion

The hydrologic model for Meenachil River was developed after the delineation of the river basin as shown in Fig. 7.

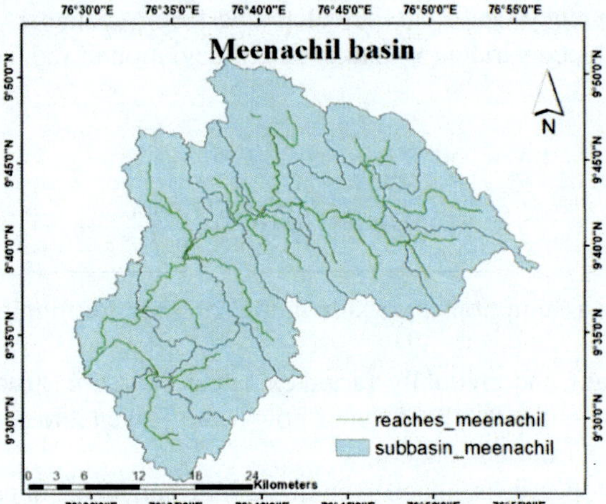

Fig. 7. Delineated Meenachil basin

Value of the parameters explained in Eq. 4 and 5 for each sub-basin is as listed in the Table 2. This lag time as shown in Table 2 was used in the SCS Unit hydrograph method for simulation. Initially, the Muskingum coefficients were assumed within their range and then their values were changed while calibrating the model.

Table 2. Calculated lagtime for each sub basin

Name	CN	S	L (ft)	Y %	Tc	Lag (min)
Subbasin-1	88	1.36	1121.01	18.16	0.10	3.72
Subbasin-10	88.20	1.34	96938.32	16.55	3.81	137.21
Subbasin-11	88.60	1.29	51040.51	11.68	2.67	96.26
Subbasin-12	88.08	1.35	66466.82	13.91	3.09	111.17
Subbasin-13	88.26	1.33	81474.63	12.65	3.78	136.25
Subbasin-14	87.55	1.42	78803.56	16.21	3.34	120.41
Subbasin-15	88.42	1.31	102076.06	11.30	4.77	171.63
Subbasin-4	87.91	1.38	88776.79	41.92	2.26	81.26
Subbasin-5	87.33	1.45	43528.14	38.97	1.35	48.71
Subbasin-6	88.02	1.36	35897.29	32.48	1.24	44.55
Subbasin-7	88.45	1.31	64486.29	10.39	3.44	123.79
Subbasin-9	89.83	1.13	72702.01	8.10	4.06	146.11

The calibration was carried out for the four monsoon months of June to September of the year 2017. The interval for simulation was one day. During the calibration process, it was observed that parameter such as Muskingum K value and lagtime had more impact while calibration.

3.1 Hydraulic Analysis Using HEC-RAS

One dimensional steady flow analysis was performed on Meenachil basin for different return periods. A total of 194 cross-sections were used to analyse the flooding behaviour of the river. The Depth of water corresponding to each return period was obtained after simulation in HEC-RAS. The obtained value was then exported to Arc-GIS to develop a flood inundation map. Flood plain maps classified based on the depth of flow for each return period was developed (Figs. 8, 9, 10, and 11). With the increase in return period, the depth of flow was also seen to be higher. It can be observed that the depth of flow rises towards the downstream sections. The area of land inundated for different return periods are as tabulated in Table 3.

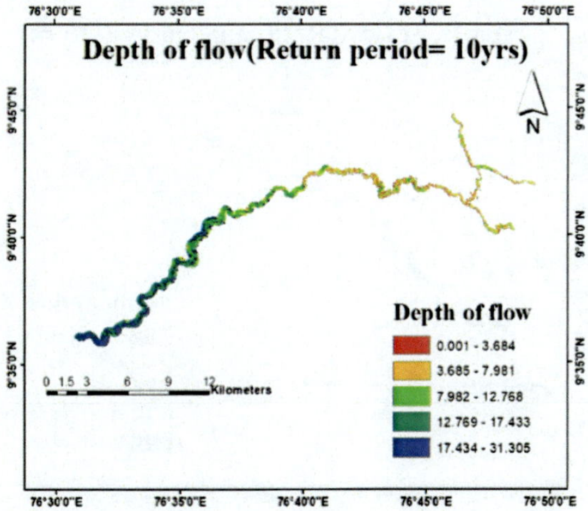

Fig. 8. Depth of flow for return period of 10 years

3.2 Flood Studies

Using the calibrated model of HEC-HMS and HEC-RAS, the analysis of 2018 flood was carried out. The peak annual discharge in cumecs was calculated using Gumbel's method and found to be 7484.468 m^3/s. This discharge value was used to undergo 1 dimensional steady flow analysis in the study area and the results obtained are discussed below. The flood plain map covered an area of 24.2465 km^2.

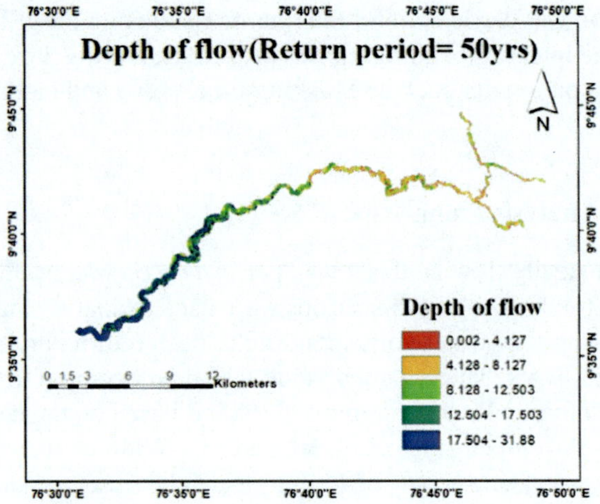

Fig. 9. Depth of flow for a return period of 50 years

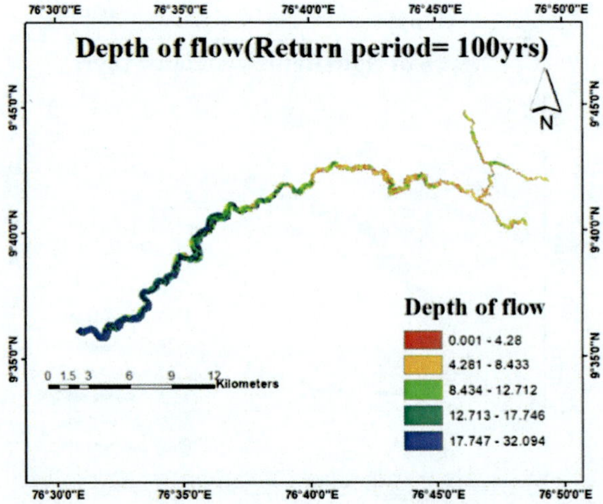

Fig. 10. Depth of flow for return period of 100 years

The LULC map (Fig. 12) was prepared using the LISS III image as mentioned in the methodology. The inundation map developed upon superimposing with the LULC map of the basin, the following classes were found to be submerged under water. The area of each classes submerged is listed in the Table 4.

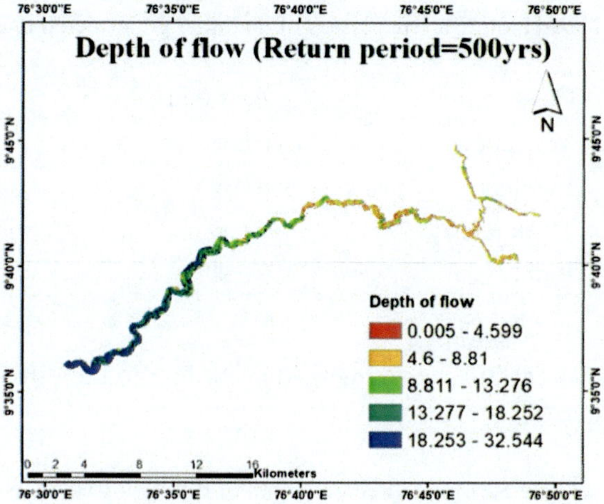

Fig. 11. Depth of flow for a return period of 500 years

Table 3. Area of land inundated for different return periods

Return period (years)	Area (km^2)
5	20.8801
10	21.2285
50	21.3857
100	21.5665
500	21.8998

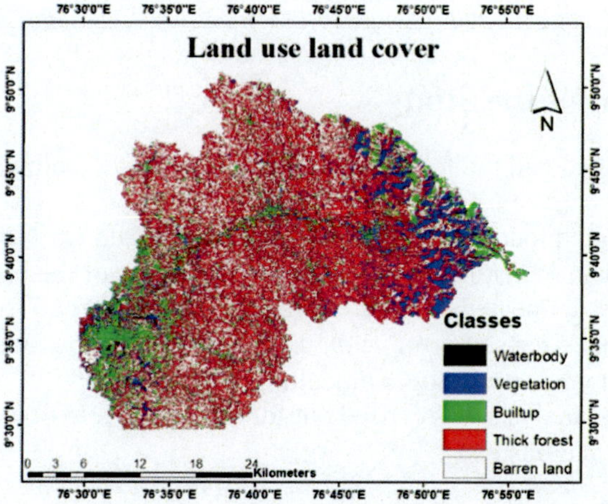

Fig. 12. LULC map of Meenachil river basin

Table 4. Area of different classes of land that got affected in 2018 flood

Class	Area (km^2)
Vegetation	6.065
Built up	4.089
Thick forest	9.490
Barren land	2.331

Similarly, the length of road submerged according to the different road classes are as listed in the Table 5.

Table 5. Flood affected length of different road classes

Road class	Submerged length (m)
Path	50.284
Primary	19474.840
Residential	11263.010
Secondary	3429.988
Service	4203.217
Tertiary	25528.530
Track	21.927
Unclassified	15631.050
Unknown	324.876

4 Limitations of the Study

The study conducted had the following limitations as listed below:

- While hydrologic modelling, the 2018 precipitation data was not used for simulation as the data had a greater anomaly compared to the rest of the historic precipitation.
- The sub-basins delineated and had lower area were merged to ease the simulation which might have caused error while calculation.Baseflow and Evapotranspiration were neglected while hydrologic modelling.
- 1 D steady flow analysis was carried out for the river while flow of water is generally 2 dimensional and unsteady.

5 Conclusion

Hydraulic-Hydrological model serve as an important tool in floodplain management studies. It can assist the urban planners and policy makers to take better action plans for the proper governance. The hydrologic model was prepared for the study area, Meenachil river basin using HEC-HMS. It consists of a basin model, Meteorological model and Control specification. Loss, Transform, Base flow and Routing methods are selected in the basin model. Soil Conservation Service (SCS) Curve number method, SCS Unit hydrograph method and Muskingum methods were the selected loss, transform and routing parameters. Baseflow and evapotranspiration were neglected. The simulated model was calibrated using 4 monsoon months of 2017. The discharge values for the year of 1990 to 2017 were simulated from the model. The peak discharge values for the return period of 5, 10, 50, 100, 500 years were found to be 757.231, 867.8569, 1111.327, 1214.255 and 1452.107 m^3/sec respectively through Gumbel's method. These discharge values were used to run the hydraulic analysis in HEC-RAS. The geometry of the river was drawn with the cross-sections in the RAS-Mapper. One dimensional steady state analysis was carried out. Flood inundation maps for different return periods are developed with details such as depth of flow and water surface elevation. Rainfall-Runoff simulation was carried out for the year of 2018 and flood inundation map was developed. The area of built-up land that got submerged in 2018 flood is 4.089 km^2 and area of vegetation submerged is 6.065 km^2. The length of different road classes submerged during 2018 flood is also calculated. The primary and tertiary roads are largely affected as 19474.84 m of primary roads and 25528.53 m of tertiary roads are submerged in 2018 flood.

References

1. Natarajan, S., Radhakrishnan, N.: Simulation of extreme event-based rainfall–runoff process of an urban catchment area using HEC-HMS. Model. Earth Syst. Environ. **5**(4), 1867–1881 (2019). https://doi.org/10.1007/s40808-019-00644-5
2. World Health Organization: WHO Guideline on the Prevention of Drowning Through Provision of Day-Care and Basic Swimming and Water Safety Skills. WHO (2021)
3. Hussain, M., et al.: Regional and sectoral assessment on climate-change in Pakistan: social norms and indigenous perceptions on climate-change adaptation and mitigation in relation to global context. J. Clean. Prod. **200**, 791–808 (2018)
4. Singh, D., Ghosh, S., Roxy, M.K., McDermid, S.: Indian summer monsoon: Extreme events, historical changes, and role of anthropogenic forcings. Wiley Interdiscipl. Rev. Climate Change **10**(2), e571 (2019)
5. Katpatal, Y.B., Patil, S.A.: Spatial analysis on impacts of mining activities leading to flood disaster in the Erai watershed India. Journal of Flood Risk Management **3**(1), 80–87 (2010)
6. CWC: Report on Kerala Floods and Solutions, Central Water Commission, Government of India (2018)
7. Borah, S.B., Sivasankar, T., Ramya, M.N.S., Raju, P.L.N.: Flood inundation mapping and monitoring in Kaziranga National Park, Assam using Sentinel-1 SAR data. Environ. Monit. Assess. **190**(9), 1–11 (2018). https://doi.org/10.1007/s10661-018-6893-y
8. Abdessamed, D., Abderrazak, B.: Coupling HEC-RAS and HEC-HMS in rainfall–runoff modeling and evaluating floodplain inundation maps in arid environments: case study of Ain Sefra city, Ksour Mountain SW of Algeria. Environ. Earth Sci. **78**(19), 1–17 (2019)
9. Dang Thanh, M., Florimond, D.: A combined hydrological and hydraulic model for food prediction in Vietnam applied to the Huong River basin as a test case study. Water **9**(11), 879 (2017). https://doi.org/10.3390/w9110879

10. Thakur, B., Parajuli, R., Kalra, A., Ahmad, S., Gupta, R.: Coupling HEC-RAS and HEC-HMS in precipitation runof modelling and evaluating food plain inundation map. In: World Environmental and Water Resources Congress, Sacramento, California, pp. 240–251. (2017). https://doi.org/10.1061/9780784480625.022
11. HEC-HMS: User Manual, Version 4.8, US Army Corps of Engineers, Hydrologic Engineering Center (2017)
12. Narayan, N., Swastik, B., Mofuzur, R., Kyle, W., Ajay, K., Sajjad, A., Ritu, G.: Flood frequency analysis using generalized extreme value distribution and foodplain mapping for Hurricane Harvey in Bufalo Bayou. Am. Soc. Civil Eng. (2018). https://doi.org/10.1061/9780784481400.034
13. Lyatkher, V.M., Proudovsky, A.M.: Hydraulic Modeling. John Wiley & Sons (2016)
14. HEC-RAS: User Manual, Davis Version 4.0, US Army Corps of Engineers, Hydrologic Engineering Center, Davis, 2008 (2021)

Analysis Based on Sediment Core Diatoms for Paleolimnological Approach

Alisha Revalia Ghassani Amir[1](✉), Tri Retnaningsih Soeprobowati[2,3,4], and Riche Hariyati[3,4]

[1] Jakarta, Indonesia
ade.alisha@gmail.com

[2] School of Postgraduate Studies, Universitas Diponegoro, Semarang, Indonesia
trsoeprobowati@live.undip.ac.id

[3] Cluster for Paleolimnology (CPalim), School of Postgraduate Studies, Universitas Diponegoro, Semarang, Indonesia

[4] Department of Biology, Faculty of Science and Mathematics, Universitas Diponegoro, Semarang, Indonesia

Abstract. Lake is a body of water surrounded by land that has an important role in human life. Galela Lake, as one of the largest freshwater sources in North Maluku, faces the threat of anthropogenic activities that impair the function of the lake ecosystem. Diatoms are used as one of the paleolimnological approaches to reconstructing environmental conditions. This study aims to examine the abundance and diversity of diatoms, as well as the status of water pollution observed through the diversity index (H′), evenness (e), dominance (D) and diatom index. Samples are taken using piston corer at the Galela Lake inlet location adjacent to the Wasi River. Using a microscope under 1000× magnification, there were 51 species of diatoms from 25 genera. The diversity index (2.10–3.21), evenness (0.69–0.92) and dominance (0.05–0.14) were categorized as medium to high, highly distributed, and low dominance. The representative diatom indices for Galela Lake are IBD, IPS, IDG, and TDI because >70% of species encountered are in accordance with the OMNIDIA database. Based on IBD and IDG, Galela Lake's inlet has good ecological status in the bottom core and gradually decreases to moderate in the upper core, while the IPS index was moderate to low. The obtained results showed that there are changes in the diatom species composition caused by anthropogenic activities around the lake that affect the change in nutrient concentration.

Keywords: Diatom · Galela Lake · Abundance · Diversity · Paleolimnology

1 Introduction

Lakes are one of the freshwater ecosystems that are important for human life to ensure the availability of water on land [1]. Galela Lake, with a surface area of approximately 390 hectares, is the largest lake on the island of Halmahera, located in the North Halmahera

S. Haddout et al. (Eds.): CCORE 2022, SPEES, pp. 131–144, 2023.
https://doi.org/10.1007/978-3-031-26967-7_10

Regency. The location of this lake becomes a potential resource for the surrounding community as drinking water, washing facilities, and land fisheries cultivation facilities [2]. However, the increase in anthropogenic activities (plantation and tourism) around Galela Lake is feared to cause changes in the quality of its water. The decline in water quality that occurred in Galela Lake was caused by increased anthropogenic activities carried out in water bodies and on land [3]. Galela Lake shows signs of eutrophication since water hyacinth plants cover about half of its surface area [2, 4]. Diatoms have a significant impact on aquatic ecosystems because their presence is often used as an environmental bioindicator [5]. The population is varied and sensitive to environmental changes, which can respond quickly and reflect changes in water quality over a long or short time [6]. Their silica cell walls are often well preserved in sediments, so diatoms are very useful for reconstructing environmental history. However, when compared to other types of microalgae, the potential of diatoms as bioindicators is underutilized [7, 8]. The application of water quality bioindicators using diatom fossils is one of the useful efforts to analyse paleolimnology by reconstructing limnological changes from the past to the future. Lake dimensions reflect the history of changes in the catchment area [9–12]. Some organisms that die and whose bodies cannot be decomposed, such as diatoms, will be stored in lake sediments which are then used to reconstruct environmental conditions [13]. The reconstruction refers to the nutritional requirements or the unique habitat of each diatom taxon, or by considering the relative abundance of species [14]. Lack of concern, diligence, professionalism in lake management results in degradation. This study aims to assess the status of water quality based on the abundance and diversity of diatoms along with the diatom index so that it can be used as a policy in the management of Galela Lake.

2 Methodology

Study Area
Galela Lake is located in Galela, North Maluku Province (Fig. 1). It is the largest lake in North Maluku with an area of 390 hectares and an altitude of 70 m above sea level. It has coordinates of 1° 49′ 6.05″ North Latitude and 127° 48′ 38.46″ East Longitude [3]. Local people know this lake by several names, including Galela Lake, Tarakani Lake and Duma Lake. Galela Lake is a closed tecto-volcanic lake with only one inflow (a temporary inlet) and no outflow (an outlet), and it is adjacent to the Wasi River, so the water retained in the lake is relatively longer than in other lakes [15, 16].

Field Work
Sediment samples were collected in October 2019. Sediment cores from surface to a depth of 90 cm were sliced at 5 cm using a piston corer at the Galela Lake's inlet adjacent to the Wasi River.

Diatom Analysis
After digestion with 10% HCL at 90 °C for two hours, sediment was rinsed with distilled water and digested with 10% H2O2 (50 ml). A neutral pH was achieved by cleansing the samples with distilled water [17]. All the samples were mounted using Naphrax with

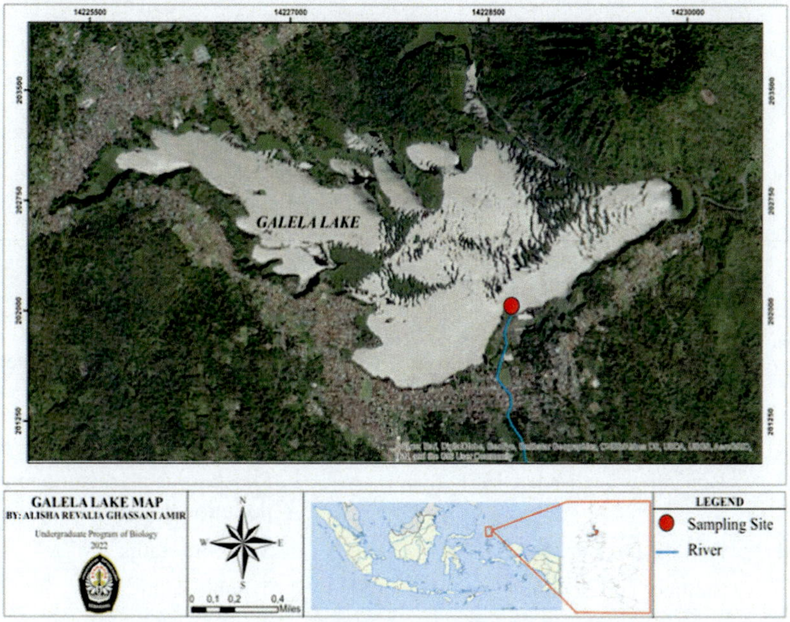

Fig. 1. Research site in Galela Lake, North Maluku

a refraction index of 1.74 and identified with a 1,000× magnification under immersion oil using a microscope, with a minimum enumeration of 300 valvae diatoms counted [18]. The diatom was identified by using identification book [19–22] and checked with AlgaBase.org [23] and diatoms.org [24].

Data Analysis

PAST 4.10 was used to compute the Shannon Wiener species diversity index (H′), as well as the species richness, evenness, and dominance indices [25]. To determine the relationship between diatom abundance and ecological variables from diatom indices, species with a relative abundance of less than 2% were removed from the calculation. The calculated diatom abundance will be utilized by OMNIDIA software to identify the environmental condition using the diatom index. This software was used to analyse the 18 diatom-based water quality indices [26], as shown in Table 1. All of the indices in the OMNIDIA software were converted to a scale of 0 to 20, indicating ecological and trophic status. The index score of 17 indicates a high ecological status and is oligotrophic; 15–17 indicates a good ecological status and is oligo-mesotrophic; 12–15 indicates a moderate ecological status and is mesotrophic; 9–12 indicates a poor ecological status and is meso-eutrophic; and 9 indicates a poor ecological status and is eutrophic (Table 2).

Table 1. Diatom indices in OMNIDIA software

Indices	Abbreviation	Correlation
CEE	Commision for Economical Community metric [27]	Physicochemical assessment
DES	Descy's pollution metric [28]	Organic pollution
EPID	Pollution metric based on diatoms [29]	Trophic status
IBD	Biological Diatom Index [30]	Phosphorus predictor
IDG	Generic Diatom Index [31]	Eutrophication and organic pollution
TDI	Trophic Diatom Index [32]	Physicochemical assessment
IDAP	Index Diatom Artois Picardie [33]	Organic pollution
IPS	Specific Pollution Sensitivity Index [34]	Eutrophication and organic pollution
Rott TI	Rott's Trophic Metric [35]	Phosphorus predictor and organic pollution
Rott SI	Rott's Saprobic Metric [36]	Saprobic status
SHE	Steinber and Schiefele's trophic metric [37]	Trophic status
TDIL	Trophic Diatom Index [32]	Trophic status
Sla	Sladecek's pollution index [38]	Organic Pollution
Lobo	Trophic-Saprobic index [39]	Eutrophication
IDP	Pampean Diatom Index [40]	Eutrophication and organic pollution
DI-CH	Swiss Diatom Index, Hurl [41]	Trophic status
WAT	Watanabe Index [42]	Saprobity
IDS/E	Louis-Leclercq Diatomic Index [43]	Eutrophication and organic pollution

Table 2. Values and trophic status for diatom indices (IBD, IDG, IPS)

Ecological status	IBD, IDG, IPS score (1–20)	Trophic status
High	>17	Oligotrophic
Good	15–17	Oligo-mesotrophic
Moderate	12–15	Mesotrophic
Poor	9–12	Meso-eutrophic
Bad	<9	Eutrophic

3 Results and Discussions

Abundance and Diversity

A total of 51 species from 25 genera were identified in diatom communities in the inlet of Galela Lake with a relative abundance of more than 2% (Fig. 2) including *Achnanthes oblongella* (2.54%), *Cymbella neocistula* (5.16%), *Diploneis calciolafrequens* (3.14%), *Encyonema latum* (3.02%), *Encyonema nicafei* (4.24%), *Encyonema sileasium* (3.50%), *Fragilaria tenera* (3.61%), *Gomphonema gracile* (4.34%), *Hantzschia amphioxys* (4.37%), *Luticola mutica* (3.12%), *Navicula leptostriata* (2.72%), *Nitzschia perminutta* (5.34%), *Pinnularia borealis* (3.46%) *and Rhopalodia rupestris* (3.15%). Compared to Afifah's research [13] that is located in the middle of Galela Lake, there are 64 species from 27 genera with the species *Aulacoseira granulate, Aulacoseira ambigua, Brachysira brebissonii, Encyonema sileasicum, Epithemia adnate, Epithemia turgida, Fragilaria biceps, Gomphonema parvulum, Navicula angusta, Navicula leptostriata, Navicula notha, Nitzschia perminutta and Ulnaria acus* are abundant. According to Lateke's research [44], cage farming sites and the Galela Lake agricultural region are home to 111 species and 35 genera. The abundant genera *Nitzschia*, *Fragilaria*, and *Luticola*, which have a wide tolerance to organic pollution and serve as indications of moderate to high levels of contamination in water, are detected in both observation sites [45].

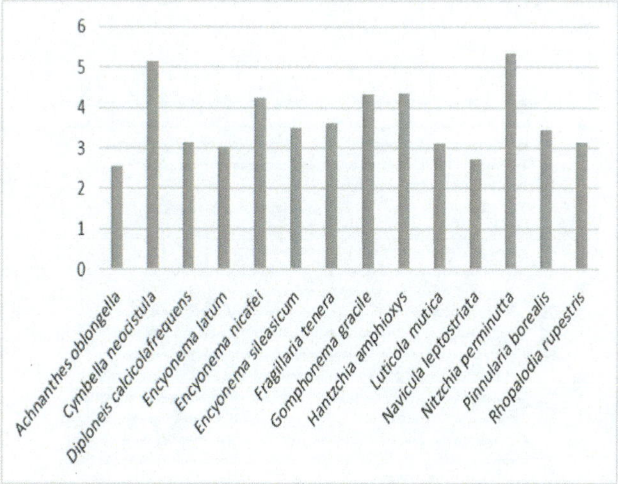

Fig. 2. Relative abundance >2% of Galela Lake

The highest abundance was *Nitzschia perminutta* (5.34%) which spread all over layers of the inlet. *Nitzschia* lives in high-nutrient environments and is able to adapt to low-light water conditions with high nutrients tolerant of pollution. They live in muddy soils and have the motility to move through sediment particles in search of nutrients [46, 47]. Followed by *C. neocistula* (5.16%), a cosmopolitan species found in mesotrophic waters with a range of pH from 6.5 to 7.5. *H. amphyioxis* with a relative abundance of

4.37%, is recorded in various ecosystems and periodically lives in dry habitats such as soil and rock crevices with aerobic environmental conditions [21, 48].

The value of the diversity index (H′) ranges from 2.10 to 3.21, (Fig. 3). The lowest diversity index at the sampling site is located in layer 40 with a value of 2.10, which indicates that the level of diversity of the layer is moderate with 9 diatom taxa found. The highest diversity index value in the 90 cm layer is 3.21 with 30 diatom taxa found. The lowest value of the Shannon-Wiener Diversity Index (H′) was at the 40 cm layer. This could be due to the abundant appearance of *Fragilaria tenera* which is proven by the dominance index value (Fig. 5) where the 40 cm layer has the highest value even though the value indicates that the dominance at that layer is low. *F. tenera* is a species that can survive in fairly high nutrient conditions and is tolerant of eutrophic waters with relatively high conductivity [49]. The 40 cm layer from the OMNIDIA software analysis has moderate mineralization with 220–600 μS/cm conductivity and oxygen with 91–96% saturation. Nitrate at the inlet location of Galela Lake from previous research [16] ranged from 0.55 to 1.0, phosphorus ranged from 0.5 to 0.7 and the highest dissolved oxygen was 6.29 mg/L. This indicates that there is an increase in nutrient enrichment activity in this layer. Mineralization and oxygen can affect the metabolism of some diatom species. Dissolved oxygen is consumed by decomposers to break down complex organic waste and create anoxic conditions in urban areas, which leads to spatial variability of dissolved oxygen, an essential element for diatom metabolism. Ions that may have an impact on the availability and uptake of nutrients by diatoms during primary production can be reflected by conductivity [50].

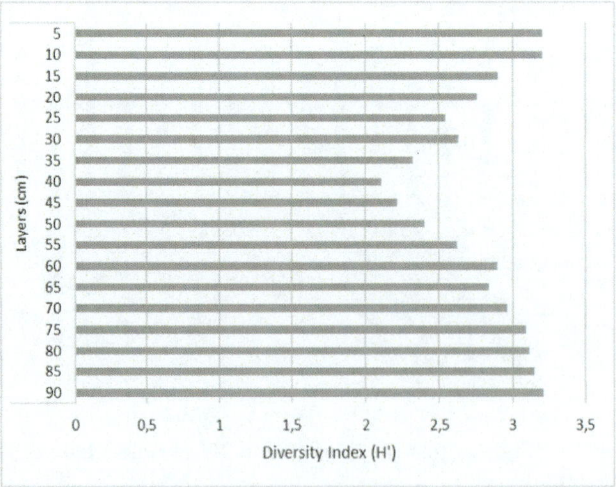

Fig. 3. Diversity index value

The result of the type evenness index (e) of this index is used to express the uneven distribution of individuals between types or whether there are certain dominant types. The observation station showed that the evenness value ranged from 0.69 to 0.92 (Fig. 4). The lowest evenness value lies in the layer of 70 cm and the highest in layer 55. The difference in evenness values that vary in each layer of Galela Lake is caused by the distribution of species in each layer, water quality, utilization, and availability of nutrients that vary from each individual. The availability of nutrients such as phosphate and nitrate, as well as the ability of each type of diatom to adapt to the existing environment [6, 51, 52].

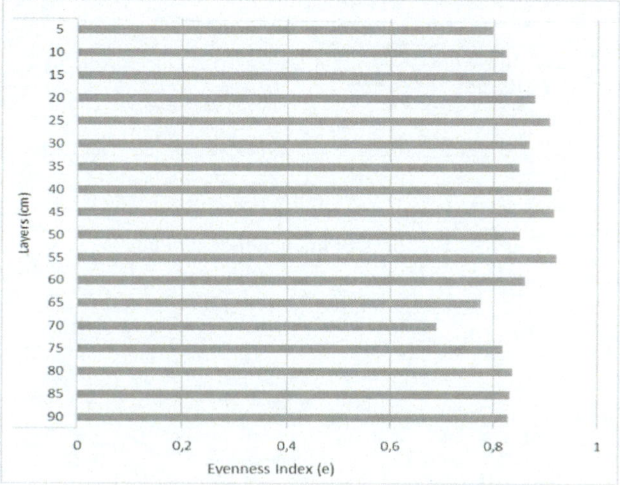

Fig. 4. Evenness index value

The index value indicates that the middle layer has a higher evenness than the top and bottom layers. This indicates the occurrence of interactions between species that complement each other, so that the high evenness value and an increase in the relative abundance of tolerant species can reduce the evenness value [53]. Sediment layers can show different characteristics due to different ages of formation [54, 55].

The highest dominance index (D) shows the highest value of 0.13 in layer 40 and the lowest in the layer of 90 cm with a value of 0.04 (Fig. 5), indicating that the diatom species that dominate the waters of Galela Lake are low. High dominance values occur in layers that are disturbed by human activities, allowing species to survive these disturbances and become the dominant species among other species [56]. The high dominance value occurs because there is a dominant species in the 40 cm layer which suppresses the growth of other species. As a result, species that cannot compete cannot thrive.

Diatom Indices

The diatom indices values of the Galela Lake study site are shown in Table 4. Diatom indices values differ depending on the species entered into the OMNIDIA calculation. The number of species entered into the calculation of the index indicates measurements that explain the ecological status. The layers variations of IBD (a), IPS (b), IDG (c), and TDI (d) values are shown in Fig. 6. IBD values range from 7.8 (poor quality) to 17.1 (high quality) (Fig. 6a). IPS and IDG values range from 10.3 (low quality) to 15.3 (good quality) and from 11.7 (low quality) to 16.3 (good quality) (Fig. 6b and 6c). The TDI ranged from 6.0 (poor quality) to 15.0 (moderate quality) (Fig. 6d).

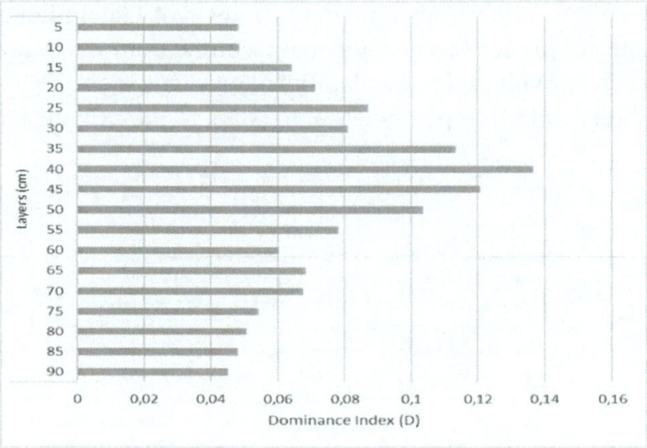

Fig. 5. Dominance index value

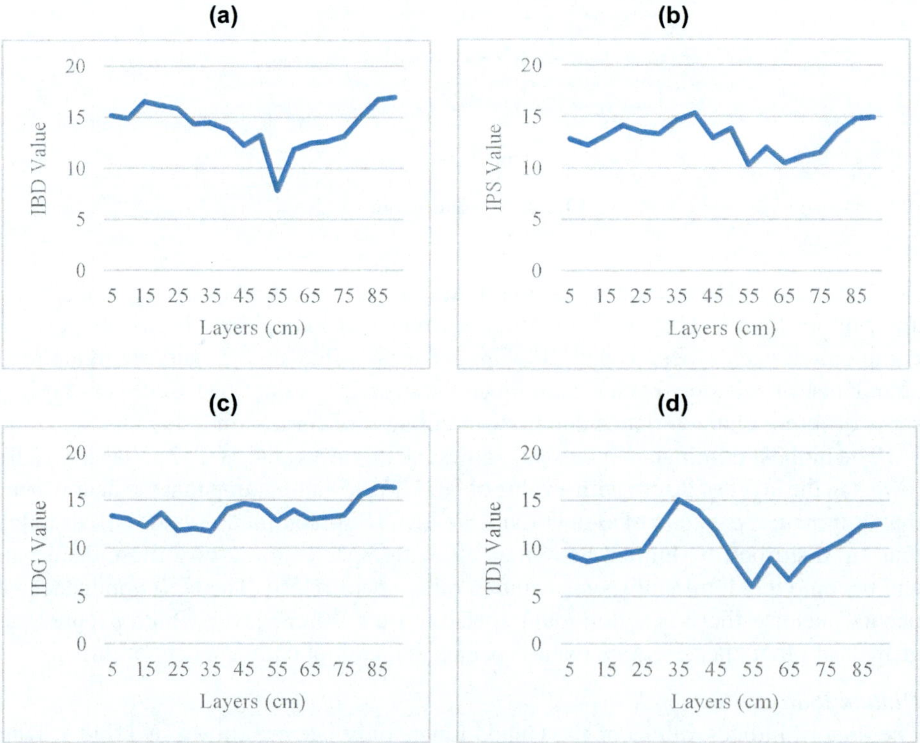

Fig. 6. Variation of the IBD (a), IPS (b), IDG (c), and TDI (d) values

Eighteen diatom indices were calculated with OMNIDIA (Table 3); however, not all the indices were encountered diatom species from Galela Lake. The number of species included in the index calculation indicated the efficiency of the resulting metric that explained the ecological status. Representative diatom indices for Galela Lake are IBD,

Table 3. Diatom indices with corresponding ecological status and percentage value

		IBD	IPS	IDG	DES	IDAP	CEE	DI-CH	TDI	SHE	Rott TI	Rott SI	Sla.	WAT	IDS/E	TDIL	EPID	Lobo	IDP
5	Index	15.1	12.8	13.3	17.0	8.9	9.9	13.6	9.2	13.8	9.9	17.0	12.7	9.0	3.68	9.1	14.3	9.6	10.5
	% Species	87.1	90.3	100	41.9	29	64.5	64.5	83.9	64.5	74.2	67.7	58.1	41.9	67.7	41.9	83.9	54.8	45.2
10	Index	14.8	12.2	13.0	16.6	9.0	9.8	13.5	8.6	14.1	8.7	16.7	12.5	9.3	3.62	8.3	13.3	11.1	11.4
	% Species	86.7	66.7	100	43.3	30	70	63.3	83.3	70	73.3	66.7	69	43.3	66.7	40	83.3	53.3	46.7
15	Index	16.5	13.1	12.2	17.0	8.6	6.4	11.7	9.0	14.1	8.4	16.8	11.9	9.8	3.54	8.4	13.7	9.5	12.0
	% Species	90.9	95.5	100	45.5	27.3	63.6	59.1	90.9	72.7	72.7	68.2	54.5	36.4	72.7	36.4	81.8	54.5	45.5
20	Index	16.1	14.1	13.6	19.0	9.6	8.4	12.2	9.4	14.5	8.8	16.0	13.3	10.5	3.77	8.5	14.6	8.0	13.8
	% Species	100	100	100	33.3	22.2	72.2	61.1	94.4	83.3	77.8	72.2	66.7	33.3	77.8	38.9	94.4	44.4	38.9
25	Index	15.8	13.5	11.8	20.0	9.1	9.9	13.9	9.7	14.2	8.6	16.0	13.4	9.6	3.93	8.7	13.7	9.4	13.8
	% Species	92.9	100	100	28.6	21.4	71.4	64.3	100	78.6	71.4	71.4	78.6	21.4	64.3	35.7	92.9	21.4	28.6
30	Index	14.3	13.3	12.0	16.9	15.9	8.5	13.0	12.4	12.4	7.9	17.3	13.7	10.5	3.63	7.7	15.8	8.8	12.1
	% Species	81.3	93.8	100	37.5	12.5	56.3	56.3	87.5	56.3	68.8	62.5	75	12.5	68.8	37.5	75	50	50
35	Index	14.4	14.6	11.7	14.0	20.0	12.1	11.0	15.0	14.5	8.2	16.5	13.9	11.3	3.64	11.8	16.3	15.0	11.5
	% Species	75	100	100	25	8.3	50	58.3	83.3	41.7	50	58.3	66.7	8.3	58.3	41.7	66.7	50	50
40	Index	13.8	15.3	14.0	10.5	0.0	11.1	11.6	13.9	12.1	11.0	18.5	14.8	0.0	3.63	12.5	17.2	10.5	10.5
	% Species	77.8	88.9	100	11.1	0	44.4	55.6	66.7	22.2	55.6	55.6	44.4	0	55.6	33.3	66.7	33.3	44.4
45	Index	12.2	12.9	14.6	5.1	5.8	9.2	5.3	11.8	7.3	5.5	16.0	13.1	9.1	3.41	8.3	12.1	11.2	9.3
	% Species	80	100	100	30	10	60	50	80	40	50	40	50	30	70	40	60	70	70
50	Index	13.2	13.8	14.3	12.4	10.1	12.6	13.8	8.7	15.3	5.5	16.3	11.7	10.8	3.33	11.6	13.4	9.3	12.0
	% Species	61.5	84.6	30.8	30.8	23.1	53.8	38.5	69.2	46.2	53.8	46.2	11.7	30.8	46.2	30.8	53.8	53.8	53.8
55	Index	7.8	10.3	12.8	9.6	8.3	7.6	10.8	6.0	11.3	5.7	14.7	12.1	8.2	2.91	8.4	10.0	9.1	9.9
	% Species	66.7	86.7	100	33.3	26.7	60	40	73.3	53.3	66.7	53.3	60	33.3	53.3	26.7	60	53.3	60
60	Index	11.8	12.0	13.9	15.2	7.5	9.4	12.6	8.9	12.8	7.7	16.0	13.7	9.6	3.4	10.3	12.7	8.0	11.9
	% Species	76.2	90.5	100	33.3	23.8	66.7	52.4	81	57.1	66.7	66.7	61.9	28.6	61.9	38.1	71.4	52.4	52.4
65	Index	12.4	10.5	13.0	16.4	4.6	8.9	10.9	6.7	11.8	8.8	16.9	12.6	9.1	3.33	9.4	12.5	7.5	11.5
	% Species	78.3	91.3	100	43.5	21.7	69.6	60.9	78.3	60.9	69.6	60.9	56.5	21.7	65.2	43.5	73.9	52.2	52.2
70	Index	12.6	11.1	13.2	14.9	5.6	9.2	8.4	8.9	10.6	8.0	16.4	13.2	9.9	3.3	10.1	14.0	6.5	10.4
	% Species	79.3	86.2	100	37.9	20.7	55.2	51.7	75.9	58.6	69	65.5	55.2	17.2	62.1	31	69	41.4	37.9
75	Index	13.1	11.5	13.4	16.0	6.7	9.5	10.7	9.8	12.1	8.0	17.0	13.1	10.3	3.46	8.7	13.8	7.0	11.4
	% Species	78.6	85.7	100	35.7	17.9	60.7	46.4	75	60.7	64.3	64.3	57.1	21.4	64.3	28.6	67.9	42.9	46.4
80	Index	15.0	13.5	15.5	16.5	9.2	11.1	10.6	10.8	12.5	9.6	17.7	14.1	10.5	3.75	7.8	15.8	7.1	11.3
	% Species	82.1	85.7	100	39.3	14.3	60.7	53.6	78.6	64.3	64.3	64.3	57.1	21.4	64.3	35.7	67.9	39.3	46.4
85	Index	16.7	14.8	16.4	17.2	10.3	12.1	11.6	12.3	13.2	11.0	18.0	14.5	10.9	3.96	8.0	16.3	7.0	11.7
	% Species	82.1	85.7	100	32.1	14.3	57.1	50	78.6	60.7	67.9	67.9	53.6	17.9	64.3	35.7	67.9	35.7	46.4
90	Index	16.9	14.9	16.3	16.5	10.9	11.8	11.3	12.5	12.8	10.9	17.9	14.4	10.9	3.94	7.7	16.0	7.7	11.6
	% Species	80	86.7	100	33.3	13.3	56.7	50	76.7	60	66.7	63.3	53.3	20	63.3	36.7	66.7	40	50

IPS, IDG, and TDI because >70% of species encountered are in accordance with the OMNIDIA database [25].

IBD value at the 90 cm layer has a good ecological status but gradually decreases as the layer gets closer to the surface. This could be due to changes in the succession of diatom sediment trends caused by anthropogenic disturbances such as industrialization and urbanization of the catchment area [57]. The disposal of household organic waste, cage cultivation, and agriculture and plantations in Galela Lake have resulted in high nitrate and phosphorus values, causing eutrophication and overgrowth of water hyacinth in the lake [16, 58]. IPS value in the bottom layer has moderate ecological status, then moderate in middle layer but has significant difference in 40 cm which showed a good ecological status. Meanwhile IDG value in the bottom layer has a good ecological status, which then decreases to poor in the middle layer but increases again to moderate in the surface layer. This could be due to the changing nutrient concentration in the lake, so that the diatom composition also changed. IBD and IPS values are correlated with each other and are significantly influenced by the concentrations of nitrate, phosphate, and chlorophyll. The IDG value is also influenced by phosphorus and chlorophyll-a [59–61]. In contrast to the other values, the TDI value peaked at the 35 cm layer. This difference could be related to the fact that the TDI index is better for temperate lakes than tropical lakes, as applied in Pasisingi's [62] research which used the TDI index to determine water quality in the upper reaches of the Cileungsi River in West Java. The application of diatom-based index values depends on the similarity of the composition of diatom species in the study area and the taxa used for each index [63]. Each taxon has an optimal value and specific tolerance for nutrients, such as phosphate and nitrite, so that their response to physical and chemical parameters varies [64].

4 Conclusion and Recommendations

There were 51 species from 25 genera were identified from the inlet of Galela Lake. The diversity index (2.10–3.27), evenness (0.67–0.92), and dominance (0.04–0.14) were categorized as medium to high, evenly distributed and low dominance. Representative diatom indices for Galela Lake are IBD, IPS, IDG, and TDI because >70% of species encountered are in accordance with the OMNIDIA database. Based on IBD and IDG, Galela Lake's inlet has a good ecological status in the bottom core and gradually decreases to moderate in the upper core. While the IPS index showed moderate to lower. The obtained results showed that there are changes in the diatom species composition caused by anthropogenic activities around the lake that affect the change in nutrient concentration, thus necessary measures should be taken by the authorities and locals to achieve sustainable lake management.

Acknowledgements. Deeply grateful to Yono, Kenanga, Yuli and Jasir for their invaluable help and support during field and laboratory work. We also thank Mirza for his help in the final identification of some diatom species.

References

1. Tyas, D.S., Soeprobowati, T.R., Jumari, J.: Water quality of Gatal Lake, Kotawaringin Lama, Central Kalimantan. J. Ecol. Eng. **22**(3), 99–110 (2021)
2. Adam, A.: Community perspective towards Eichhornia in Galela Lake, South Galela, North Halmahera. Archipelago **1**(1) (2020)
3. Soeprobowati, T.R., Jumari, J., Saraswati, T.R., Suhry, H.C., Gell, P.: Land-use changes concerning the riparian vegetation in Galela Lake, North Maluku, Indonesia. Ecol. Eng. **170**, 106368 (2021)
4. Coetzee, J.A., Hill, M.P.: The role of eutrophication in the biological control of water hyacinth, Eichhornia crassipes, in South Africa. Biocontrol **57**(2), 247–261 (2011). https://doi.org/10.1007/s10526-011-9426-y
5. Komalasari, E., Khairuddin, K., Lalu, J.: The Diatom community in Maluk Coastal water in West Sumbawa. In: Proceedings of the 1st ACCESS: Advances in Social Science, Education and Humanities Research, vol. 465 (2019)
6. Soeprobowati, T.R.: Variabilitas keanekaragaman dan distribusi vertikal diatom Danau Rawa Pening. J. Sci. Math. **19**(3), 65–70 (2011)
7. Bueno, G., et al.: Automated diatom classification (Part A): handcrafted feature approaches. Appl. Sci. **7**(8), 753 (2017)
8. Saxena, A., Tiwari, A., Kaushik, R., Iqbal, H.M.N., Roberto, P.: Diatoms recovery from wastewater: overview from an ecological and economic perspective. J. Water Process Eng. **39**, 101705 (2021)
9. Adrian, R., et al.: Lakes as sentinels of climate change. Limnol. Oceanogr. **54**(6), 2283–2297 (2009)
10. Battarbee, R.W., Bennion, H., Gell, P., Rose, N.: Human impacts on lacustrine ecosystems. In: The Sage Handbook of Environmental Change. Sage (2011)
11. Di, B., Liu, D., Wang, Y., Dong, Z., Li, X., Shi, Y.: Diatom and silicoflagellate assemblages in modern surface sediments associated with human activity: a case study in Sishili Bay, China. Ecol. Ind. **24**, 23–30 (2013)
12. Zhang, W., et al.: Lake sediment records on climate change and human activities in the Xingyun Lake catchment, SW China. PLoS ONE **9**(7), e102167 (2014)
13. Afifah, J.: Penentuan Jumlah Minimum Frustula Dalam Identifikasi Diatom Danau Galela, Halmahera Utara, Maluku Utara. Thesis, Universitas Diponegoro, Semarang (2021)
14. Soeprobowati, T.R., Suedy, S.W.A., Lubis, A.A., Gell, P.: Diatom assemblage in the 24 cm upper sediment associated with human activities in Lake Warna Dieng Plateau Indonesia. Environ. Tech. Innov. **10**, 314–323 (2018)
15. Abdullah, A.: Studi Bioekologi Ikan Air Tawar di Danau Galela Kabupaten Halmahera Provinsi Maluku Utara. IPB Press, Bogor (2005)
16. Suhry, H.C., Soeprobowati, T.R., Saraswati, T., Jumari, J.: Kualitas Air dan Indeks Pencemaran danau Galela. J. Il. Ling. **18**(2), 236–241 (2020)
17. Soeprobowati, T.R., Suedy, S.W., Hadiyanto: Diatom and water quality of Telaga Warna Dieng, Java, Indonesia. In: IOP Proceeding: Earth and Environmental Sciences, vol. 55 (2016)
18. Soeprobowati, T.R., Djalal, T.S., Sutikno, H.S., Gell, P.: The minimum number of valvas for diatom identification in RawaPening Lake, Central Java. Biotropia **23**(2), 96–104 (2016)
19. Gell, P., Sonneman, J.A., Reid, M.A., Illman, M.A., Sincock, J.: An illustrated key to common diatom genera from Southern Australia. Cooperation Research Centre for Freshwater Ecology, Identification Guide No. 26, Thurgoona, NSW, 65 (1999)
20. Sonneman, J.A., Sincock, A., Fluin, J., Gell, P.: An illustrated guide to common stream diatom species from temperate Australia. Cooperative Research Centre for Freshwater Ecology, Hurgoona, N.S.W, 11 (1999)

21. Taylor, J.C., Harding, W.R., Archibald, C.: An Illustrated Guide to Some Common Diatom Species from South Africa. Water Research Commission, Gezina (2007)
22. Krammer, K., Lange-Bertalot, H.: Bacillariophyceae 4th part achnanthaceae, kritische erganzungen zu navicula (lineolatae), gomphonema gesamtliteraturverzeichnis teil 1-4 [second revised edition] [with "ergänzungen und revisionen" by h. langebertalot in: h. ettl et al; Freshwater Flora of Central Europe. spektrumakademischer Verlag Heidelberg. 2, 4, 1] (2004)
23. Guiry, M.D., Guiry, G.M.: Worldwide electronic publication, National University of Ireland, Galway. AlgaeBase (2022). www.algaebase.org. Accessed 3 Aug 2022
24. Spaulding, S.A., et al.: Diatoms of North America. Diatoms (2022). https://diatoms.org. Accessed 3 Aug 2022
25. Hammer, O.: Paleontological statistics software package for education and data analysis OMNIDIA version 4.10 (2022). http://download.canadiancontent.net/PAST.html
26. Lecointe, C., Coste, M., Prygiel, J.: 'Omnidia': software for taxonomy, calculation of diatom indices and inventories management. Hydrobiologia **269**, 509–513 (1993). https://doi.org/10.1007/BF00028048
27. Descy, J.P., Coste, M.A.: Test of methods for assessing water quality based on diatoms. Verh. Intern. Limnol. **24**, 2112 (1991)
28. Descy, J.P.: A new approach to water quality estimation using diatoms. Nova Hedwig. Beih. **64**, 305 (1979)
29. Dell'uomo, A.: Assessment of water quality of an Apennine river as a pilot study for diatom-based monitoring of Italian watercourses. In: Whitton, B.A., Rott, E. (eds.) The Use of Algae for Monitoring Rivers II, p. 65. Institutf¨ur Botanik Universität Innsbruck, Innsbruck (1996)
30. Lenoir, A., Coste, M.: Development of a practical diatom index of overall water quality applicable to the French National Water Board network. In: Whitton, B.A., Rott, E. (eds.) The Use of Algae for Monitoring Rivers II. Institutf¨ur Botanik Universität Innsbruck, Schmitten (1996)
31. Coste, M., Ayphassorho, H.: Study of the water quality of the Artois-Piccardie Basin using benthic diatom communities (application of diatomic indices). Reprot CEMAGREF, Bordeaux, Artois-Picardie Water Agerncy, Douai, France, p. 227 (1991)
32. Kelly, M.G., Whitton, B.A.: The trophic diatom index: a new index for monitoring eutrophication in rivers. J. Appl. Phycol. **7**, 433–444 (1995). https://doi.org/10.1007/BF00003802
33. Prygiel, J., Coste, M.: Recent trends in monitoring French rivers using algae, especially diatoms. In: Whitton, B.A., Rott, E. (eds.). Use of Algae for Monitoring Rivers II, p. 87. Institutfür Botanik Universität Innsbruck, Schmitten (1996)
34. Cemagref. Etude des méthodes biologiques d'appréciation quantitative de la qualité des eaux in Bassin Rhône-Mediterrannée-Corse, eds Q. E. Rapport and A. F. Lyon. CSIRO, Lyon, p. 218 (1982)
35. Rott, E., et al.: Lists of indications for growth algae. Federal Ministry of Agriculture and Forestry, Vienna, p. 248 (1999)
36. Rott, E., Hofmann, G., Pall, K., Pfister, P., Pipp, E.: Lists of indications for algae in Austrian rivers, Part 1: Saprobial indication waer management cadastre. Federal Ministry of Agriculture and Forestry, Vienna, p. 73 (1997)
37. Schiefele, S., Schreiner, C.: Use of diatoms for monitoring nutrient enrichment acidification and impact salts in Germany and Austria. In: Whitton, B.A., Rott, E., Friedrich, G. (eds.) Use of Algae for Monitoring Rivers, p. 193. Institutf¨ur Botanik Universität Innsbruck, Schmitten (1999)
38. Sládeček, V.: Diatoms as indicators of organic pollution. Acta Hydrochim. Hydrobiol. **14**, 555 (1986)

39. Lobo, E.A., Heinrich, C.G., Schuch, M., Wetzel, C.E., Ector, L.: Diatoms as bioindicators in rivers. In: Necchi, O. (ed.) River Algae, pp. 245–271. Springer, Cham (2016). https://doi.org/10.1007/978-3-319-31984-1_11
40. Gomez, N., Licursi, M.: The Pampean Diatom Index (IDP) for assessment of rivers and streams in Argentina. Aquatic Ecol. **35**, 173–181 (2001). https://doi.org/10.1023/A:1011415209445
41. Hürlimann, J., Niederhauser, P.: Method of analysis and assessment of water courses in Switzerland, Diatom, level R region. Federal Office for the Environment, Forests and Lanscape. OFEFP, Berne **111**, 81 (2002)
42. Watanabe, T., Asai, K., Houki, A.: Numerical water quality monitoring of organic pollution using diatom assemblages. Sci. Tot. Env. **55**(1) (1986)
43. Leclerq, L., Maquet, B.: Two new chemical and diatomic running water quality indices: comparison with various existing indices. Cahiers de Biologie Mar. **303** (1987)
44. Lateke, S.: Biostratigrafi Diatom Danau Galela Halmahera Utara-Maluku Utara. Dissertation, Universitas Diponegoro, Semarang (2022)
45. Krammer, K., Lange-Bertalot, H.: Süßwasserflora von Mitteleuropa, Bd 2/2. Bacillariophyceae. 2. Teil: Bacillariaceae, Epithemiaceae, Surirellaceae. Gustav Fisher Verlag, Berlin (1988)
46. van der Grinten, E., Janssen, A.P.H., de Mutsert, K., Barranguet, C., Admiraal, W.: Temperature and light-dependent performance of the cyanobacterium Leptolyngbya Foveolarum and the Diatom Nitzschia Perminuta in mixed biofilms. Hydrobiologia **548**, 267–278 (2005). https://doi.org/10.1007/s10750-005-5324-6
47. Klamt, A.M., Hu, K., Huang, L., Chen, X., Liu, X., Chen, G.: An extreme drought event homogenises the diatom composition of two shallow lakes in southwest China. Ecol. Ind. **108**, 105662 (2020)
48. Sonneman, J.A., et al.: An illustrated guide to common stream diatom species from temperate Australia. Cooperative Centre for Freshwater Ecology, Albury (2000)
49. Mercurio, A.L., Querijero, B.L., Ching, J.A.: Phytoplankton in aquaculture and non-aquaculture sites of Taal Lake, Batangas, Philippines. J. Exp. Biol. Agric. Sci. **4**, 66–73 (2016)
50. Saros, J., Fritz, S.C.: Nutrients as a link between ionic concentration/composition and diatom distributions in saline lakes. J. Paleolimnol. **23**(4), 449–453 (2000). https://doi.org/10.1023/A:1008186431492
51. Baker, V., Savage, R.: Development of a wetland monitoring program for headwater wetlands in North Carolina. NC Department of Environment and Natural Resources, NC Division of Water Quality, EPA, North Carolina (2008)
52. Sirait, M., Rahmatia, F., Pattulloh, P.: Komparasi indeks keanekaragaman dan indeks dominansi fitoplankton di sungai ciliwung jakarta. J. Kel. **11**(1), 75–79 (2018)
53. Passy, S.I., Bottin, M., Soininen, J., Hillebrand, H.: Environmental filtering and taxonomic relatedness underlie the species richness–evenness relationship. Hydrobiologia **787**(1), 243–253 (2016). https://doi.org/10.1007/s10750-016-2968-3
54. Soeprobowati, T.R., Suedy, S.W.: Komunitas diatom pada ekosistem mangrove pantai utara jawa tengah. J. Sci. Math. **18**(3), 94–102 (2010)
55. Rahmawati, D.: E-modulStratigrafi: Teknik Geologi. Universitas Mulawarman, Samarinda (2021)
56. Ulfah, M., Fajri, S.N., Nasir, M., Hamsah, K., Purnawan, S.: Diversity, evenness and dominance index fish in Krueng Raya Water, Aceh Besar. In: IOP Conference Series: Earth and Environmental Science, vol. 348, p. 012074 (2019)
57. Mckay, A.W., et al.: Diatom succession trends in recent sediments from Lake Baikal and their relation to atmospheric pollution and to climate change. Philos. Trans. Roy. Soc. B Biol. Sci. **353**(1371), 1011–1055 (1998)

58. Zubcov, E., Laurentia, U., Antoaneta, E., Nina, B., Natalia, B.: Influence of nutrient substances on phytoplankton from Prut River. Ann. "Dunarea De Jos" University of Galati **32**, 68–72 (2009)
59. Gbadebo, A.M., Dimowo, O.B., Taiwo, A.M., Omoniyi, I.T.: Application of diatom-based indices in river quality assessment: a case study of lower Ogun river (Abeokuta, Southwestern Nigeria) using epilithic diatoms, vol. 3, no. 1. IntechOpen (2019)
60. Yang, J., et al.: Applicability of benthic diatom indices combined with water quality valuation for Dish Lake from Nanjishan nature reserve, Lake Poyang. Water **12**, 2732 (2020)
61. Acs, E., Reskone, N.M., Szabo, K., Taba, G., Kiss, K.T.: Application of epiphytic diatoms in water quality monitoring of Lake Velence – recommendations and assignments. Acta Bot. Hung. **47**, 211–223 (2005)
62. Pasisingi, N., Pratiwi, N.T.M., Krisanti, M.: The use of trophic diatom index to determine water quality in the upstream of cileungsi river. West Java. Ind. Fish. Res. J. **20**(1), 1–14 (2014)
63. Tan, X., Sheldon, F., Bunn, S.E., Zhang, Q.: Using diatom indices for water quality assessment in a subtropical river, China. Environ. Sci. Pollut. Res. **20**(6), 4164–4175 (2012). https://doi.org/10.1007/s11356-012-1343-9
64. Harding, W.R., Archibald, C.G.M., Taylor, J.C.: The relevance of diatoms for water quality assessment in South Africa: a position paper. Water SA **31**(1), 41–46 (2005)

Impact of Agro-Chemicals Exposure on the Human Health and Environment

Shanta Kumari[1](✉) and Chetan Chauhan[2]

[1] Department of Economics, Eternal University, Baru Sahib, Sirmour 173101, Himachal Pradesh, India
shantakaushalchauhan@gmail.com

[2] Department of Chemistry, Sardar Patel University, Mandi 175001, Himachal Pradesh, India

Abstract. In modern agricultural technology agrochemicals are inevitable inputs but its indiscriminate use in agriculture has serious repercussions on the farmers' health and the environmental. The Kullu district in Himachal Pradesh was chosen purposively for the study. The secondary data has been used for the year 2006 and primary data was collected for the year 2017 from 100 farmers by using pre-tested schedule. From the last decades, intensification of the agriculture has led to extensive use of agro-chemicals by the farmers especially in cash crops like apple and vegetable crops. Most of the farmers using agro-chemicals without using the protective equipment and have its direct exposure. The study concluded that agro-chemicals exposure is more in recent time than the earlier. The practice of indiscriminate agro-chemicals is due to inferior quality, resistance developed by the pest and the effect of climate change. Therefore, there is a need to enhance and provide depth knowledge to the farmers to strengthen their understanding regarding the use of agro-chemicals which have direct or indirect impact on human health and environment.

Keyword: Agrochemicals · Environmental health · Protective Equipment

1 Introduction

Agro-chemical is a common term for the various chemical products, such as fertilizer, fungicide, insecticide, hormone or soil treatment that protect the crops and also and increases its production. Since 1948, India is using pesticides and its production started in 1952 [1]. In India, the use of pesticides (0.3 kg/ha) is comparatively less than other developing countries like Sri Lanka (0.4 kg/ha) and Indonesia (0.8 kg/ha) [2]. Although agricultural productivity increased by the use of agro-chemical, but it more use affects adversely on human health and environment resulting in loss of working efficiency and agro-biodiversity. Pesticides are toxic chemicals and its exposure creates risk to users. The level of risk increases to those users who are often illiterate, ill-trained and not using appropriate protective equipment while handling the agrochemicals. This increases more numbers of bad effects of pesticides. Therefore, these bad effects in human being incurring the biggest cost of environmental which paid by the society for the their use

S. Haddout et al. (Eds.): CCORE 2022, SPEES, pp. 145–153, 2023.
https://doi.org/10.1007/978-3-031-26967-7_11

[3, 4]. The agro-chemicals exposure have morbidity effects. These have been studies by numerous researchers through the blood samples [5–12]. Many long-term chronic effects created by the use of Pesticide exposure are beyond the scope of this study. The present study depend on self-reported symptoms, may not reveal the real prevailing facts. There are few survey based studies on related issues in Kullu district of Himachal Pradesh done earlier [3, 13] but comparative study is not found. Therefore, the present study attempted to fill the research gap regarding the agro-chemicals use and its influence on the farmer's health and its environment.

Objective of the Study
To find out the impact of the agro-chemicals on the health of human and environment of the study area during the two point of time.

2 Methodology

The selection of Kullu block from the Kullu district (Fig. 1) was done purposively. From the selected block, the list of panchayats was organized. Among the Panchayat's list, Jallugran panchayat was selected randomly. Later the list of villages falling in the Jallugran panchayat was prepared and half of the villages selected randomly. The proportion allocation method was used for the selection of 100 households from the selected villages. The data was taken from those farmers who were doing spray (vegetable and apple crops) most of the time from many years in each household.

The study is based on secondary [3] as well as primary data. The primary data has been collected for the agricultural year 2017–18 through a pre-tested schedule from the sample households on the aspects like pesticide exposure, farmers characteristics (age, height & weight) year & adoption of IPM, rate of spraying pesticides, spray time & number of spray, use of protective equipment and symptoms of pesticides exposure were used for the collection of data. To construct Body Mass Index (BMI), the weight and height of the respondents was noted from each household. This has been calculated by the ratio of weight (kg) to the Height^2 (m). Data were analyzed through percentages and tabular method used to present the results.

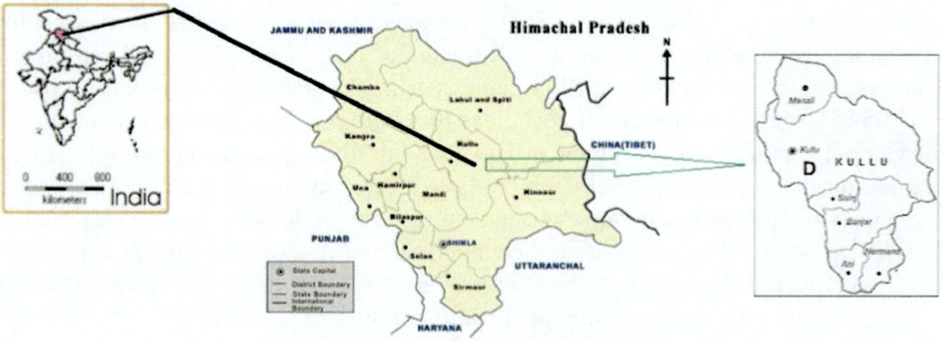

Fig. 1. Map of the study area

3 Results

In Table 1 result of the study expressed that average age, weight and height of a person who did spray were 40.33 years, 57.64 kg and 1.61 m in 2006, respectively. While in 2017, these features were 41.01 years, 60.24 kg and 1.66 m.

Table 1. Physical characteristics of respondents

(Person/Farm)

Particulars	2006	2017
Age (Years)	40.33	41.01
Weight (Kg)	57.64	60.24
Height (m)	1.61	1.66

Source: Kumari&Sharma, 2014 and field survey, 2017

The more no. of farmers were having normal weight in both the years (Table 2). The problem of underweight and overweight have been increased in 2017.

Table 2. Physical characteristics of the respondents in terms of body mass Index

(Person/Farm)

Particulars	2006	2017
Under weight	9	17
Normal	79	67
Overweight	10	15
Obese	2	1

Source : Kumari & Sharma, 2014 and field survey, 2017

In 2006, from long time farmers were using pesticides (Table 3). For example, fifty percent of them were using pesticides for the last 25 to 30 years. Whereas, in 2017 in the range of 30–35 years same percent was engaged. The table also exposes that frequency of spraying was a little higher in later years in comparison to earlier. It was also pointed that 28 (2006) and 45 (2017) percent of the households had adopted integrated pest management.

Table 3. Years and frequency of using agro-chemicals and adoption of IPM

	No. of Respondents	
Duration of Agro-chemical application	**2006**	**2007**
10-15	1	0
15-20	25	0
20-25	50	0
25-30	24	8
30-35	0	50
>35	0	42
Frequency (No. of Spray)	7	9
Adoption of IPM		
Yes	28	45
No	72	55

Source: Kumari & Sharma, 2014 and field survey, 2017

The result on the different aspects of pesticide use like no. of spray, type of pesticides and time of spray, etc. are given in Table 4. The nine to ten sprays in 2006 done by the 41 per cent of the households while 6–8 sprays followed by around 48 per cent of households. On the other hand, in 2017, less than 50% of farmers were doing spray of pesticides from 6 to 8 times, while in the range of 9–11 times 55 per cent of households were doing pesticides spray. Further, in both years all the sample households were using insecticides and fungicides for spray, except during flowering. More number of households were found to using spray for colour in 2017 in comparison to 2006.

All the farmers were making partial use of kit in 2017, whereas in 2006 it was slight less (70%) (Table 5). At the time of spray, all the farmers were wearing old clothing. Only 22 per cent of the farmers in 2006 and 90% in 2017 used the polythene to cover nose and mouth. Farmers feel uncomfortable while bearing the clothes during the spray which was the main the reason for using less protective gears. Less than half of the farmers in 2006 also reported that they were not interested in using protective measures.

Table 4. Pesticide use and its impact on pollinators

	Per cent of Respondents	
Particulars	2006	2017
No. of Spray		
1-2	0	0
3-5	11	0
6-8	48	45
9-11	41	55
Type of pesticides used		
Insecticide	100	100
Fungicides	100	100
Timing of spray		
Before flowering	100	100
During Flowering	72	82
During Fruiting	100	100
Before harvesting (For Colour)	61	85
Do pesticides kill insect?		
Yes	90	100
No	2	0
Don't know	8	0

Source: Kumari&Sharma, 2014 and field survey, 2017

Table 5. Use of protective measures while spraying pesticides

(Per cent of Respondents)

Particulars	2006	2017
User	70	90
Non-user	30	0
Use of kit		
Fully	0	0
Partially	70	100
No	30	0
Measures used		
Gloves	12	60
Boots	40	60
Old clothes	100	100
Cover nose and mouth with Polythene	22	90
Reasons for non-use		
Not interested	45	0
Uncomfortable	90	100
Unnecessary	30	10

Source:Kumari&Sharma, 2014 and field survey, 2017

Majority of the farmers reported to have experienced acute illnesses due to pesticides exposure (Table 6). In 2006, most of them opined that they had experienced eye irritation (86 per cent) followed by 81% who reportedly experienced fatigue, 66% skin irritation, 59 per cent head ache and back pain, 56 per cent vomiting, 22 per cent dizziness and 1 per cent eye discharge. In 2017, all the respondents reported irritation in eye, fatigue (90%), headache & skin irritation (85%), dizziness (55%) and eye flu (15%). The 59 per cent household responded that they had back pain and 56 per cent of the households were having the symptoms of vomiting. In 2006 (84%) and 2017 (45%) those farmers who had suffered due to the illness of pesticides exposure availed the clinic facility. In 2006, 16 per cent farmers and 55 per cent farmers had not availed clinic facilities after the illness due to pesticides exposure (2017).

Table 6. Multiple responses of the farmers regarding agro-chemicals poisoning: symptoms of pesticides

(Per cent of Respondents)

Symptoms	2006	2017
Eye Irritation	86	100
Headache	59	85
Dizziness	22	55
Vomit	56	56
Back Pain	59	59
Skin Irritation	66	85
Eye flu	1	15
Fatigue	81	90
Availing clinic		
Yes	84	45
No	16	55

Source: Kumari&Sharma, 2014 and field survey, 2017

4 Discussion

From the various studies it has been found that the indiscriminate use of agro-chemicals causing more adverse effects on the health of the human beings and the surrounding environment. The results of the study revealed that problems related to health such as underweight and overweight were found more in 2017 than the former. This might be due to the injudicious use of agro-chemicals. The farmers of the study area were engaged in the cultivation of the commercial crops rather than cereal crops which enhanced the use of agro-chemicals. The frequency of pesticides spray were found higher in later period in comparison to former in high value cash crops. All the farmers were sprayed the pesticides before the flowering, fruiting and after fruiting in both time periods. Whereas, during the time of flowering, most of the farmers were sprayed the pesticides in 2017 than earlier. Almost all the farmers responded that in both the time period spray of pesticides killed insect and bees which could be due to excess use of agro-chemicals. With respect to the adoption of IPM, 45% of the farmers reported its use in 2017 and 28% in 2006. The farmers were not adopted IPM could be due to inferior quality of the agro-chemicals.

Multiple responses of the farmers regarding agro-chemicals poisoning and their symptoms were calculated on the basis of self-reported data so there could be underestimation of health issues due to pesticide use. The usage of less protective measures was found among the pesticides applicators. In 2006, more than two-third of the farmers partially used the kit provided by the horticulture department while in 2017 all farmers were using the provided kit. The kit provided with spray pump was used by very less farmers as the farmers do not feel comfortable while spraying the pesticides in the hilly terrain of the study area. The own protective measures taken by most of the farmers were the use of old clothes, gloves and shoes in both time period. Very few farmers were covering their nose and mouth with polythene while spraying in 2006 and most of the farmers were applied the same practice in 2017.

5 Conclusion

It has been concluded from the study that the frequency and doses of agro-chemicals used by the farmers has been increased in 2017 in comparison to 2006 which creating more health problems and environmental issue like loss of pollinator. It could be due to climate change, use of sub-standard agro-chemical formulations and resistance developed by the pest against the agro-chemicals. It has been also noticed that farmers were not fully protecting themselves while applying the agro-chemicals. Therefore, intensive training to understand the impact of agro-chemicals through case study on the health and environment by the experts of their respective area should be done. To encourage the farmers to use full kit of protective measures, there is a need to manufacture farmer friendly protective gears to make them comfortable during handling and application of the agro-chemicals. Further, encouragement should be created among the farmer to enhance use organic farming/ natural farming which will ultimately help to bring the sustainability. This is the pressing need of the hour and also concerned scientists should do the study on the same respondents by taking and diagnosing their blood sample and soil sample to know about the actual facts of agro-chemicals on human health and environment. It will be vital to find out actual status of the study in question for taking further suitable action.

References

1. Bhardwaj, T., Sharma, J.P.: Impact of pesticides application in agricultural industry: an Indian scenario. Int. J. Agric. Food Sci. Technol. **4**(8), 817–822 (2013)
2. Biswas, S.K., Rahman, S., Kobir, S.M.A., Ferdous, T., Banu, N.A.: A review on impact of agrochemicals on human health and environment: Bangladesh perspective. Plant Environ. Dev. **3**(2), 31–35 (2014)
3. Kumari, S., Sharma, H.: The impact of pesticides on Farmer's health: a case study of fruit bowl of Himachal Pradesh. Int. J. Sci. Res. **3**, 44–48 (2014)
4. Wilson, C.: Private costs and the relation between pesticide exposure and ill health: evidence from Sri Lanka. Environ. Econ. Policy Stud. **5**(3), 213–227 (2002). https://doi.org/10.1007/BF03353922

5. Rola, A.C., Pingali, P.L.: Pesticides, Rice Productivity, and Farmers' Health: An Economic Assessment, pp 10–108. International Rice Research Institute, Washington, DC (1993). http://books.irri.org/971220037X_content.pdf
6. Antle, J.M., Pingali, P.L.: Pesticides, productivity, and farmer health: a Philippine case study. Am. J. Agr. Econ. **76**(3), 418–430 (1994)
7. Pingali, P.L., Marquez, C.B., Palis, F.G.: Pesticides and Philippine rice farmer health: a medical and economic analysis. Am. J. Agr. Econ. **769**(3), 587–592 (1994)
8. Crissman, C.C., Antle, J.M., Capalbo, S.M.: Economics, Environmental and Health Tradeoffs in Agriculture: Pesticides and the Sustainability of Andean Potato Production, pp 10–28. International Potato Centre, Kluwer Academic Publishers, Dordrecht (1998)
9. Antle, J.M., Cole, D.C., Crissman, C.C.: Further evidence on pesticides, productivity and farmer health: Potato production in Ecuador. Agric. Econ. **18**, 199–207 (1998)
10. Cole, D.C., Carpio, F.J., Julian, L.N.: Economic burden of illness from pesticide poisonings in highland Ecuador. Panam. Am. Rev. Public Health **8**(3), 196–201 (1998)
11. Dasgupta, S.C., Meisner, D., Wheeler, D., Xuyen, K., Lam, N.T.: Pesticide poisoning of farm workers–implications of blood test results from Vietnam. Int. J. Hyg. Environ. Health **210**(2), 121–132 (2007). https://doi.org/10.1016/j.ijheh.2006.08.006
12. Sharma, S., Kumar, K., Bhargava, S., Jamwal, V.S., Sharma, A., Singh, R.: Data in support of poisoning related mortalities from southern Himachal Pradesh. Data Brief **12**, 493–498 (2017)
13. Kumari, S.: High value cash crops agriculture in Himachal Pradesh: a study in documentation and valuation of environmental costs. Ph.D. thesis, CSK HPKV, Palampur, Himachal Pradesh (2007)

Climate Change Impact on Agriculture of Almaty Region, Kazakhstan

Zhansaya Bolatova(✉)

Department of Agribusiness and Consulting, Kazakh National Agrarian Research University, Almaty 040009, Kazakhstan
72311jan@gmail.com

Abstract. The global processes of climate change under the influence of anthropogenic factors entail extreme and almost irreversible consequences. Climate change affects agricultural production and its productivity throughout the world. Agriculture, which is one of the major sources of greenhouse gas emissions, can play an important role in mitigating the effects of climate change. Kazakhstan is Central Asia's most important supplier of grain and grain products. Kazakhstan also plans to increase the number of livestock for 100% domestic meat supply, but the issue rests on the quality and availability of pastures affected by climate change. Analyzes showed that under the conditions of the expected climate in 2030, the average yield of spring wheat in Kazakhstan regions (Akmola, Aktobe, West Kazakhstan, Karaganda, Kostanay, Pavlodar) will be decreased and in the conditions of 2050 – 51–80% will be a decrease of production. This means that while maintaining the existing stage of farming culture, the yield of spring wheat by 2030 will decrease to 13–37%, which will lead to a reduction in the area by 23.86% of the total area of crops in 2019. Objectives of the review: to analyze the processes of climate change and to study the economic impacts of climate change on agriculture in Kazakhstan. For article have been used different literature as IPCC, WMO, WTO, FAO, UNEP, UNFCCC, UNDP, IMF, WB, OECD, KAZHYDROMET, IRRI, Committee of the Statistics of Kazakhstan, Turkish State Meteorologic Service reports etc.

Keywords: Climate change · Agriculture production · Impacts of climate changes · Greenhouse gas emissions · Yields

1 Introduction

Kazakhstan is the 9th biggest country in the world and has a macroclimatic region. Kazakhstan is simultaneously located in Eastern Europe and Central Asia, it is washed by the Aral and Caspian Seas. Winters in this area are cold and dry, while summers are dry and hot. Almost half of the territory of Kazakhstan is semi-deserts and deserts. Mountain ranges are located in the western part. Water resources are also scarce, which is a geographic location. If we talk about vegetation, you can most often find wormwood, feather grass and shrubs in the steppe, which have an increased resistance to long droughts. Soil is of great importance for farming. Most of the territory falls on brown and chestnut soils,

S. Haddout et al. (Eds.): CCORE 2022, SPEES, pp. 154–163, 2023.
https://doi.org/10.1007/978-3-031-26967-7_12

as well as on black soil. Corrosive soil and gray soils are also present. Kazakhstan has difficult topographic peculiarities and a transition zone for diverse pressure systems, air masses. Kazakhstan has 2 main industries that are traditionally represented: crop production is the basis of agriculture. Crops such as corn, oats, millet, barley, buckwheat, rice and spring wheat are grown on the territory of the country. Considerable areas are also occupied by oilseeds and sugar beets, grapes, melons, apples, potatoes, flax and cotton; animal husbandry in the country is represented by cattle breeding, as well as goats, pigs, camels, horses and sheep [7, 12–14].

Among all regions of Kazakhstan, relatively homogeneous in terms of climatic characteristics, the Almaty region is characterized by violent diversity. The region, stretching between Lake Balkhash and the mountains of the Zailiysky Alatau, combines five climatic zones. Here it is equally possible to develop agriculture and cattle breeding. Here, rivers alternate with sands, and mountains, whose sharp peaks rest directly on the sky, flow into green plains and golden steppes. It combines five natural and climatic zones, which cause a wide variety of ecosystems. There are two reserves and five natural parks on the territory of the region. However, such paintings risk sinking into oblivion if climate risk management methods are not applied at the state level as soon as possible [12–14].

The number of greenhouse gas emissions showed the influence of the modern stage of the country's development. Due to climate change, agriculture and food companies are facing significant challenges, having to respond quickly to changing environmental and regulatory requirements. Agriculture affects climate by greenhouse gases and climate affects crop yields through climate change. Climate change affects the quantity and quality of food, as well as food security. The country's resources should be directed to climate change mitigation and train qualified employees [17–20].

According to available data, climate change is impacting agriculture, including food security, and thus threatens to spread hunger, which is becoming increasingly difficult to eradicate. Because it was the 150-year phase of rapid economic growth and the resulting increase in greenhouse gas emissions that led to a global increase in temperature by an average of 1 °C compared to the pre-industrial period. At current rates, average global warming between 2030 and 2050 is expected to likely reach 1.5 °C. At the same time, it is worth noting that in the agricultural sector, climate change and the impact on yields are more important in agricultural development policy, which also includes prices, consumption, food security, economic well-being, etc. [1–5].

Some authors in their works included the impact of climate in partial and general equilibrium models. Understanding how climate change affects agricultural welfare is important both for policy and because the simple integrated assessment models (IAMs) used to calculate the social cost of carbon (SCC) use damage functions that parameterize changes in economic welfare depending on temperature. According to estimates based on various climate models, future climate change will reduce the annual harvest of pulses in many countries. Many researchers predict significant future climate change costs in the absence of unprecedented adaptation. At the present stage of development, the Republic of Kazakhstan is experiencing a significant impact of the changing climate on the regions such as economic and agricultural resources. At the same time, the main problems of climate change are a decrease in economic growth, deterioration in food security and a decrease in labor productivity in agriculture [4–6, 8–10, 15].

According to the statistics on crop production in Kazakhstan, between 2004–2005 and 2012–2013, there was an upward trend in yields, but from 2006 to 2011 it is worth noting that yields were low. In Kazakhstan, from 2006 to 2011 there was a drought, and as a result, a low yield of field crops was observed. It is also worth noting that over the past 70 years, the annual and seasonal surface air temperatures in Kazakhstan have increased. At the same time, the average annual temperature in the country increases by 0.27 °C every 10 years. Literature sources show that the greatest warming was in autumn by −0.32 °C, while winter and spring temperatures increase more slowly by −0.29 °C every 10 years. The minimum warming in summer was 0.20 °C in 10 years. The contribution of this trend to the total dispersion of the average annual temperature is 37%. The seasonal contribution varies from 6 to 27% [8, 9, 17, 18]. The global processes of climate change under the influence of anthropogenic factors entail extreme and almost irreversible consequences. Climate change affects agricultural production and its productivity throughout the world. Increases in population in the future are demand from diet changes. It can only be met through further productivity increases from the expansion of agricultural land, which is extremely limited. The increase in temperature also affected precipitation and, as a result, influenced the development of the country's economy, including rising fuel prices and agricultural productivity [20]. Currently, the vast majority of climate scientists support the concept of global warming, naming the anthropogenic factor as one of the causes of warming. Even taking into account the ongoing policy to reduce CO2 emissions, there is a 90% chance of an increase in global temperature by 2100 by 2.0–4.9 °C. The warming of the global climate will manifest itself in different ways, but it will be most noticeable in the middle and high latitudes due to the melting of ice [3–5, 19].

2 Materials and Methods

The objectives of the review are to analyse the processes of climate change and to study the economic impacts of climate change on agriculture in Kazakhstan. For article have been used different literature as IPCC, WMO, WTO, FAO, UNEP, UNFCCC, UNDP, IMF, WB, OECD, KAZHYDROMET, IRRI, Committee of the Statistics of Kazakhstan reports, etc. Materials have been used from different literature and statistics, which showed by % of years.

3 Results and Discussion

Kazakhstan areas have largest sown areas are located in 4 regions: Kostanay (more than 5.1 million hectares), Akmola (more than 4.8 million hectares), North Kazakhstan (more than 4.3 million hectares) and Almaty region (0.9–1.3 million ha). At the same time, 66% of the sown area is located in the three northern regions of Kazakhstan (Fig. 1). It should be noted the diversity of climatic and natural conditions in the country. Agriculture in the Almaty region is developing in conditions of high air temperature in the foothills. If artificial irrigation is right here, then a good harvest of tobacco, sugar beets, rice and cotton can be harvested. Also in this production, there is quite a profitable occupation in viticulture.

Fig. 1. Map of agriculture production distribution [9]

Kazakhstan's agriculture is experiencing a certain stagnation. This situation is characterized as incomplete and irrational use of human and natural resources. The main problem is the still incomplete transition to a modern farm form, which greatly complicates the control over tax deductions. It should also be noted that there is insufficient investment in the agricultural sector. Most of the problems exist in the meat and dairy industry, which has led to the forced import of products to satisfy the needs of consumers. Another important problem that needs to be solved as soon as possible is the lack of space where the harvested crop can be stored. The Almaty region is most exposed to threats associated with climate change. Among all regions of Kazakhstan, relatively homogeneous in terms of climatic characteristics, the Almaty region is characterized by violent diversity.

Climate change has been affected by land degradation and desertification in Kazakhstan. During a drought occurs increased evaporation and the supply of moisture in the soil becomes insufficient, which affects the country's economy and ecology. Drought in the country is very common (Fig. 2) and has been observed 8 times over the past 51 years (from 1975 to 2012), and also tends to recur. After the winter season, land degradation occurs, it is aggravated by natural disasters such as floods, mudflows, landslides and fires [12, 28, 30]. According to the forecasts of some literature, the frequency of mudflows will increase 10 times and pose a threat to 15 regions of Kazakhstan, including the Almaty region.

Thus, as a result of the study of the impact of climate change on the agriculture of Kazakhstan, measures are proposed to adapt to climate change in the agro-industrial complex of the regions. Mainly for measures to adapt agriculture to the consequences

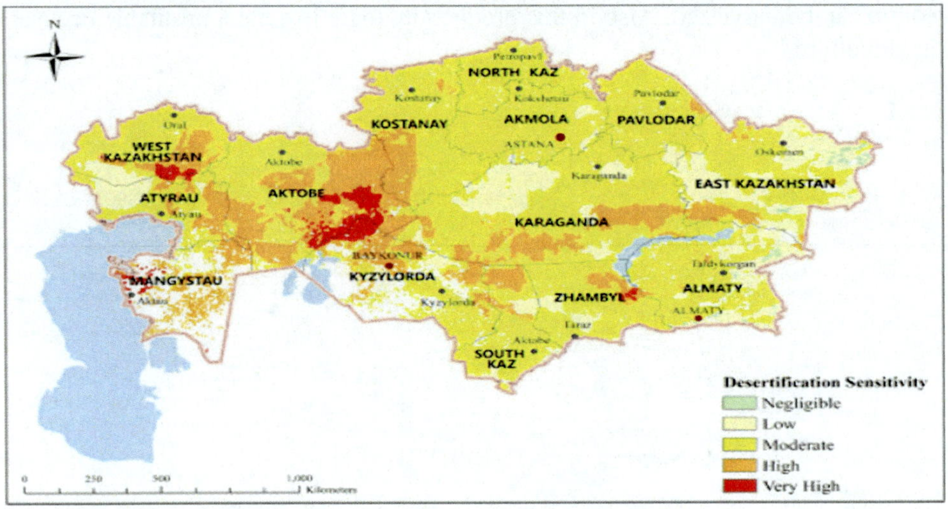

Fig. 2. Desertification sensitivity of Kazakhstan [30].

of climate warming in the following areas: new technologies; weather forecast; state technical provision of agriculture; state scientific and educational support for agricultural producers; state information support of manufacturers; a climate change insurance system in agriculture should be made for producers. The solution to these problems also depends on the collective and interdisciplinary efforts and cooperation of public authorities and the scientific community.

Adaptation to the harmful effects of climate change encourages the transformation of the interaction between society and the ecosystem. Kazakhstan is a trans bordered country of water resources with China, Kyrgyzstan, Uzbekistan and Russia. The domestic demand for water resources have been increased in the country due to demographic growth and accelerating economic stability. Kazakhstan cannot provide sufficient water resources for irrigation, due to the policy of neighboring countries. Seven of the country's water management basins are transboundary and highly have been dependent on neighboring countries' water management policies. The country's strategically important crops depend on climate and irrigation systems that have long since fallen into disrepair due to poor management. Public investment in research on the adaptation of the agro-industrial complex to climate change should be allocated.

The state should move to monitor and calibrate climate and agricultural factors. Nowadays, the arable soil of the country suffers from water shortages and extreme climate. To cope with the challenges posed by climate change, it is necessary to develop low-carbon renewable energy sources. The Almaty region is most exposed to threats associated with climate change.

The yield of wheat is higher than the average republican level (10.9 c/ha) in Almaty (17.2 c/ha). In the Almaty region, climate change has already led to such consequences as a change in the water regime of mountain rivers, degradation of glaciers, depletion of water resources, and an increase in abnormal weather events: extreme heat, droughts, dust storms, etc. This region is also at increased risk of extreme hydrometeorological situations, such as avalanches, mudflows, floods and so on. Potential losses from the destructive force of erosion (mudflows) in the Almaty region, according to experts, could amount to 1 million 125 thousand US dollars. The importance of the ecological situation in the Almaty region is also because both agriculture and cattle breeding are actively developing here. At the same time, as you know, agriculture consumes a significant amount of water resources. But the trouble is that the main volume of water resources is located in Balkhash and Alakol, the waters of which are unsuitable for irrigation. At the same time, irrational water use, the lack of new irrigation technologies, and an outdated and inefficient irrigation system are observed [7, 21, 22, 24–27].

The coefficient of use of water resources is 0.95, while the environmentally acceptable limit is 0.65. The ecological state of the region is very tense. Main problems: soil degradation, shifting sands, depletion of biodiversity. For example, the area of forests in the Dzungarian and Zailiyskiy Alatau has decreased by 24%. Ecosystem degradation is 70%. The vegetation is changing and pasture lands subject to wind erosion are high. Changes in temperature and precipitation, a decrease in moisture, and an increase in the aridity of the climate will lead to the gradual death of ecosystems. In addition, the region, rich in natural diversity, is also subject to a large number of disasters: earthquakes, floods, mudflows, landslides, etc. [7, 25].

It should be noted that the entire region of Central Asia is under the destructive influence of global climate change. This threatens not only weather influences such as freezing rain, but the impact will also all areas of life.

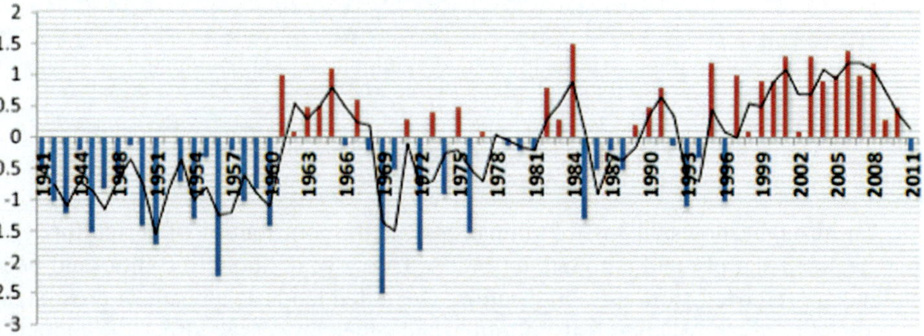

Fig. 3. Temperature abnormalities 1941–2011 [12].

An increase in temperature affects an increase in the level of seas, ocean acidification and the economy of every country. The relationship between anthropogenic forcing and weather conditions impacts on populations' adaptation to changing climatic conditions.

Kazakhstan suffers significant economic losses as a result of a 1 °C increase in average annual temperature, which will slow down economic growth by 0.9%. For a low-income developing country with an average annual temperature of 25 °C, the effect of a 1 °C rise in temperature is even greater: a 1.2% drop (Fig. 3). The economy is estimated to be hit hard by up to a 20% rise in temperatures. Over the past 70 years, annual and seasonal surface air temperatures have increased. The average annual increase of 0.27 °C over 10 years. The impact of climate change is shown in Fig. 3. The greatest warming was in autumn up to 0.32 °C for 10 years. Winter and spring temperatures rise slightly more slowly, at 0.29 °C every 10 years. The minimum warming in summer was 0.20 °C for 10 years. The total average annual temperature dispersion was 37%. The seasonal impact ranges 6–27% [8–11, 15, 20, 28].

GDP in agriculture has fallen to 10% in recent years due to agricultural production in rainfed lands, which makes production resilient to climate change. Under the conditions of the expected climate in 2030, the average yield of spring wheat in seven studied Kazakhstan regions will be decreased. In the conditions of 2050 - 51–80% of production will be decreased. This means that while maintaining the existing stage of farming culture, the yield of spring wheat by 2030 will decrease to 13–37%, which will lead to a reduction in the area by 23.86% of the total area of crops in 2019 [7, 25, 27].

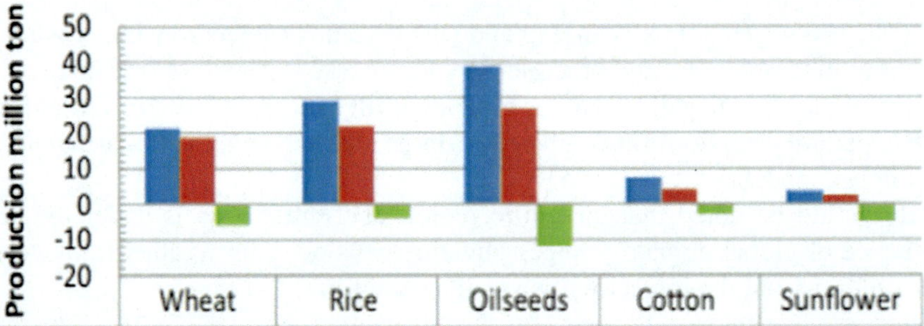

	Wheat	Rice	Oilseeds	Cotton	Sunflower
Normal	21.4	29	38.7	7.4	3.7
Drought	18.7	22	26.9	4.3	2.5
Change %	-6	-4	-12	-3	-5

Fig. 4. Impacts of drought crop production [8]

The drought shows the loss of grain crops and the contribution of agriculture to GDP has declined sharply and amounts to 24.2%. Every year, about 250 thousand hectares of land are removed from the soil rotation. Degraded lands as a result of erosion, salinization, waterlogging, chemical pollution and other processes affect the reduction in crop yields. The area of degraded pastures reaches 60 million hectares and 15 million hectares are taken out of agricultural use. The gross harvest of grains and leguminous crops was high in 2007 and 2009 at up to 20 million tons. Due to the dry years of 2008 - 2010, the gross harvest decreased significantly and amounted to only 12.2 million tons (9.6 million tons of wheat). The total cultivated area now stands at more than 21 million hectares, with more than 75% of this area planted with cereals (Fig. 4).

The agricultural sector in Kazakhstan produces strategically important crops that are completely dependent on irrigation systems and due to poor water management, unsuitable agricultural practices [22].

In Kazakhstan, adaptation and mitigation impact plans are funded annually. US$727 million have been expanded to adaptation, US$396 million to mitigation, and the overlap amounted to US$26 million in 2013–2014 (Fig. 5). The drought impact decreases the harvest of products and it's affected the market and the price [8, 23, 28]. Agriculture

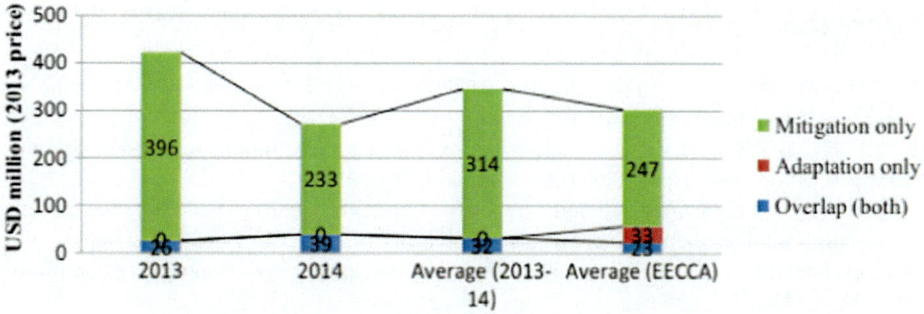

Fig. 5. Climate change mitigation and adaptation [12].

is highly vulnerable to climate change, frequent droughts and water shortages affecting domestic production. To cope with climate change issues, the current approach is to focus on the development of low-carbon, renewable energy sources. Several key strategies and concepts have been adopted to outline strategic directions for national climate change mitigation and adaptation actions.

4 Conclusion

Climate change affects through drought and flooding with rising and falling temperatures that affect crop yields and crop production. According to a study of some researchers, shown that the yield of cereals has decreased by 10%. Almaty region adaptation of climate change and its negative impacts raises a significant transformation of society and the natural ecosystem. There is an increase in domestic consumption of water resources due to population growth and the acceleration of economic development. With the rapid impact of climate change, security first results in the development of adaptation of agriculture. Despite the presence of fertile land to take into account the lack of water and harsh climatic conditions of Almaty region, Kazakhstan.

Asseng et al. (2015) found that climate change is expected to reduce global wheat production, but IslyamiA., et al. (2020) research shows that Kazakhstan could witness gains in some areas and under some conditions and losses in others [28, 29]. Research by many scientists shows that climate change's impacts have different effects, as humidity increases in Norwegian regions and droughts in CIS countries. Sirkku J. (2017) published that corn production is increased from the effects of climate change, but Wang D. et al. (2022) found that in Kazakhstan production of cereals decreased [19, 26]. It means that every region has specific effects from climate change, which can increase and decrease welfare.

Some suggestions: Kazakhstan for climate change needs to make a new adaptation political program and diversify cultivation methods and sustainable irrigation management, preparation at universities and at the state level of lectures or courses for farmers who are affected by the impact of climate change and agricultural insurance should include the impact of climate change, develop methods for using and storing rain and snow water for irrigation.

References

1. Nelson, G.C., et al.: Agriculture and climate change in global scenarios: why don't the models agree? Agric. Econ. **45**, 85–101 (2014)
2. Nelson, G.C., et al.: Climate change effects on agriculture: economic responses to biophysical shocks. Proc. Natl. Acad. Sci. **111**(9), 3274–3279 (2014)
3. Wiebe, K., et al.: Climate change impacts on agriculture in 2050 under a range of plausible socioeconomic and emissions scenarios. Environ. Res. Lett. **10**, 2–10 (2015)
4. Stevanović, M., et al.: The impact of high-end climate change on agricultural welfare. Sci. Adv. **2**, e1501452 (2016)
5. Stevanović, M., et al.: Mitigation strategies for greenhouse gas emissions from agriculture and land-use change: consequences for food prices. Environ. Sci. Technol. **51**(3), 65–74 (2017)
6. Маршалл, Б., Эмерик, К.: Адаптация к изменениюклимата: данныесельскогохозяйст ва США. Американскийэкономическийжурнал: Экономическаяполитика **8**(3), 106–140 (2016)
7. Worldbank: Climate-smartagriculture (2021). https://www.worldbank.org/en/topic/climate-smart-agriculture
8. OECD: Climate-smart agriculture. Agriculture and climate change: Towards sustainable, productive and climate-friendly agricultural systems (2016). https://www.oecd.org/agriculture/ministerial/background/notes/4_background_note.pdf
9. FAO: Climate-smart agriculture (2021). http://www.fao.org/climate-smart-agriculture/en/
10. EPA: Sources of Greenhouse Gas Emissions (2021). https://www.epa.gov/ghgemissions/Sources-greenhouse-gas-emissions#agriculture
11. EPA: Global Emissions by Economic Sector (2021). https://www.epa.gov/ghgemissions/global-greenhouse-gas-emissions-data#Sector
12. Ministry of Environmental Protection of the Republic of Kazakhstan, RSE Kazhydromet: Kazakhstan Annual Climate Change Monitoring Bulletin, Astana (2020)
13. The Organization for Economic Cooperation and Development: Country report. Financing Climate Action in Kazakhstan (2016)
14. The State of Food and Agriculture: Drought characteristics and management in Central Asia and Turkey (2017)
15. IPCC: Climate change. WGII Sixth Assessment Report (2022). www.ipcc.ch/report/ar6/wg2/downloads/report/IPCC_AR6_WGII_FinalDraft_FullReport.pdf
16. Bureau of National Statistics: Agency for Strategic Planning and Reforms of the Republic of Kazakhstan. Agriculture, Forestry and Fisheries in the Republic (2021)
17. БолатоваЖ, Б., БулхаироваЖ, С., КульшигашоваМ, К.: Экономикаизмененияклима тарегионовКазахстана и еговлиянияна АПК. XIМеждународнаянаучно-практическая конференция «Современныепроблемыцивилизации и устойчивогоразвития в информа ционномобществе, pp. 333–339 pp, УДК 332.2.021.012.32 (2022)
18. БолатоваЖ, Б., Енгиндениз, С.: Экономикаизмененияклимата в сельскомхозяйствеКа захстанаЭкономика и экологиятерриториальныхобразований **5**(2), 25–35 (2021). ISSN: 2413-1474
19. Sirkku, J.: Climate change transformations in Nordic agriculture? J. Rural. Stud. **51**, 28–36 (2017)
20. Bolatova, Z., Engindeniz, S.: The economics of climate change in agriculture: a review on Kazakhstan and Turkey. In: Mattas, K., Kievit, H., van Dijk, G., Baourakis, G., Zopounidis, C. (eds.) Sustainable Food Chains and Ecosystems. CM, pp. 1–15. Springer, Cham (2020). https://doi.org/10.1007/978-3-030-39609-1_1
21. NSA: Основныесоциально-экономическиепоказателиРеспубликиКазахстан (Main Socio-Economic Indicators of the Republic of Kazakhstan). National Statistical Agency,

Astana (2017). http://stat.gov.kz/faces/homePage/homeDinamika.pokazateli. Accessed 20 Dec 2017

22. Robinson, S., et al.: The international model for policy analysis of agricultural commodities and trade (IMPACT): model description for version 3. IFPRI Discussion Paper 1483. IFPRI. Washington, DC, 26 November 2015. https://doi.org/10.2139/ssrn.2741234
23. Strokov, A., Mirzabaev, A., Alexey, Y., Kiselev, S.: Irrigation development as a measure to combat climate change risks. In: Proceedings of the 22nd International Sustainable Development Research Society Conference, ISDRS, vol. 1, pp. 662–679 (2016)
24. Wandel, J., Kozbagarova, B.: Kazakhstan: economic transformation and autocratic power. Mercatus Policy Series, Country Brief 4. Mercatus Center, George Mason University, Arlington (2009). https://doi.org/10.2139/ssrn.1434522
25. World Bank: World Development Indicators Database. The World Bank, Washington, DC (2017). https://data.worldbank.org
26. Wang, D., et al.: Impact of climate change on food security in Kazakhstan. Agriculture **12**, 1087 (2022). https://doi.org/10.3390/agriculture12081087
27. World Bank, Asian Development Bank: Climate risk country profile – Kazakhstan (2021). https://climateknowledgeportal.worldbank.org/sites/default/files/2021-08/15834-WB_Kazakhstan%20Country%20Profile-WEB.pdf
28. Islyami, A., et al.: Impact of climate change on agriculture in Kazakhstan. Silk Road A J. Eurasian Dev. **2**(1), 66–88 (2020). https://doi.org/10.16997/srjed.19
29. Asseng, S., et al.: Rising temperatures reduce global wheat production. Nat. Clim. Change **5**, 143–147 (2015). https://doi.org/10.1038/nclimate2470
30. Hua, Y., Hana, Y., Zhanga, Y.: Land desertification and its influencing factors in Kazakhstan. J. Arid Environ. **180**, 104203 (2020)

Why is Correct Agricultural Water Management Necessarily a Prerequisites in Water Shortage Regions?

Bilal Acar(✉), Nuh Uğurlu, Sena Afacan, Emrah Gülen, Nasuh Açık, Erdal Kökdere, and Yusuf Taştan

Faculty of Agriculture, Department of Farm Buildings and Irrigation, University of Selçuk, Konya, Turkey
{biacar,nugurlu}@selcuk.edu.tr

Abstract. This study was carried out to investigate the significance of proper water management in water-stressed agricultural areas such as Turkey's Konya Basin. For that purpose, agro as well as the water potential of Konya plain was analyzed in regard to water management. As results, water resources particularly groundwater, are not used sustainably since there is dramatic groundwater depletion in most parts of the region. The reasons are the widening of cultivated lands in the favor of crops having grand water consumption and areas brought into irrigation with no care. In accordance with our previous research findings in our region, the following solutions could be addressed: crop pattern redesigned in accordance with current water resources for example crop pattern of cereals could be increased, landowners producing low water consuming crops must be subsidized, sprinkler or drip irrigation systems should be used more, water charges should be volume-basis, and farmers should be trained about correct and deficit irrigation with visual materials.

Keywords: Climate change · Agro-water management · Water use in agriculture · Water savings

1 Introduction

Water is a vital important input in agriculture and maximum crop growth, as well as crop yields, can be obtained in the conditions of adequate water availability within the crop root zone depth during whole vegetation cycles [1]. Possibly one of the most important issues in water-poor ecologies is successful agricultural water management [2, 3]. Water and fertilizer are known as two necessary inputs in agricultural production. Improvement of the soil's physical properties is very important for optimal water and nutrient distributions in the rooting environment [4]. Water scarcity is a worldwide crisis, so plenty of countries have focused on productive utilization of water resources for example sprinkler and drip irrigation known as water-saving modern technologies have been used at 1.683 million hectares (51%), and 0.660 million hectares (20%) areas, respectively in Italy [5].

S. Haddout et al. (Eds.): CCORE 2022, SPEES, pp. 164–170, 2023.
https://doi.org/10.1007/978-3-031-26967-7_13

There is no doubt that sprinkler and drip irrigation systems have resulted in notable water saving, particularly in arid and semi-arid regions, including Turkey. It was stated [6] that low water costs are one of the most important factors for abundant water use in agriculture.

Water productivity in irrigation is based on two factors: water delivery performance and field water application efficacy. The age and quality of the water distribution networks are highly effective in water delivery performance [7].

The main crops are cereals, sugar beet, maize, sunflower, dry beans, potatoes and carrot in the Konya plain. Among those, sugar beet, maize, alfalfa, and carrot are highly water-consuming crops.

The annual available surface water resources are around 1.867 billion m3 and 1.567 billion m3/year have been used under current in Konya closed basin. There is no over water utilization from the surface water bodies. However, although annual available ground water resources are around 2.00 billion m3, 3.50 billion m3/year have been used under current so groundwater potential of Konya Closed Basin have not been used sustainably.

In the world general, 95% of fresh water is obtained from groundwater since groundwater is the perfect quality because porous rock in an aquifer sieves water and holds the suspended particles as well as bacteria. The consumption of groundwater increases day by day in some countries such as the United State of America. Around 67% of groundwater is used for irrigation, such as in Texas, Arizona, and California. The increase in groundwater use results in some problems likes aquifer depletion (in case of faster groundwater withdrawal than recharging by precipitation), subsidence of groundwater by withdrawal of groundwater and accessing of salt water into fresh water in coastal areas [8]. Overwater consumption is a serious ecological problem, mainly in water shortage environments. For example, water level reductions were recorded at 10 m and 30 m for NW and SE parts of Iran, respectively, between 2010 and 2014 [9].

The main aims of the current study are to analyze the agricultural potential of the Konya plain, and propose necessary recommendations for sustainable agricultural water management for such kinds of regions.

2 Methodology

The data used in this paper were obtained from the some documents or reports relevant to our region called Konya plain. In addition, the data about the water productivity under deficit conditions were based on the our field research conducted at Konya province of Turkey in most and some from the different worldwide environments such as Togo, and Ethiopia where they are suffering from the water scarcity problems.

Konya plain, mostly focusing on in our studies, is characterized as semi-arid climate in accordance of long-term average annual rainfall of 323 mm with 1285 mm evaporation. Yet despite the rich farmland in the Konya plain, the region's water resources make up only 2.5% of Turkey. Negative effects of climate change on agricultural output in the region are significant. On the other hand, although Konya plain has great farming areas the water resources are only 2.5% of Turkey. Overuse of the groundwater reservoir in the Konya Closed Basin, amounts to roughly 1.5 billion m^3 per year, causing an annual decline in water level of about 1.0 m. Land area in Konya plain is about 4 × 106 hectares, which is 5.2% of the total land area in Turkey and 8.15% of the total farming area. About 9.22%, 4.0%, and 1.5% of Turkey are dedicated to the cultivation of field crops, vegetables, and fruits, respectively, in the province of Konya.

3 Results and Discussion

It is common knowledge that in low-water areas, agricultural operations are profoundly affected by the total amount and distribution of rainfall. It is crucially critical to have even distribution of rainfall throughout the crop vegetation cycle. As a result, in the semiarid Konya province, summer crops benefit greatly from light rain throughout the crucial growth periods (early April to late September).

A meteorological drought has witnessed by a period of below-average rainfall relative to the average rainfall amount for that period in Middle Anatolia region of Turkey. A rise in the severity of the drought has inevitably led to a decrease in agricultural output, resulted by a decrease in water availability. Water stress has occurred in the soil profile during agricultural drought period. Hydrological drought then observed by an indication of low water levels in or the entire depletion of water supplies as a result of excessive evaporation or transpiration. In brief, a low precipitation level has caused drought environments [10].

Irrigation has many functions, the most important of which are increasing farmers' income, ensuring that water reaches crops with as little waste as possible, and minimizing the negative environmental impacts of water use (such as sinkhole development due to excessive pumping from aquifers). Adopting the proper irrigation strategy is crucial for long-term agro-water sustainability.

Seventy percent of the world's fresh water is used in agriculture [4], and nearly eighty percent in the Konya plain, Turkey.

Summer crops in the Konya region cannot be grown successfully without irrigation in order to generate a sustainable income from farming. For this reason, irrigation is one area where water should be used very efficiently. There were two key factors limiting the long-term viability of our region's water supply. At start, there was no particular strategy behind the expansion of irrigated land. Second, as a result of the maximum profits they bring in, farmers have planted more and more water-intensive crops like maize on their property. In most of the Konya basin, groundwater supplies have been depleted due to over use. If this is the case, rising irrigation energy costs and decreasing groundwater levels will drive up groundwater pumping prices. Because of this problem, farmers whose land is irrigated are seeing a lower net return. Over-abstraction of groundwater can be avoided if crop patterns can be defined in accordance with the amount of water resources in the region [11]. Spreading out the use of high water productivity irrigation methods like sprinkler or drip irrigation was found another viable option for sustainable irrigation in water-poor locations [12]. One solution to the water shortage problem was to increase water productivity and incorporate marginal waters [3]. Variables like crop and soil types, climate, and irrigation management all have affected how well an irrigation system functions [13]. While both surface and drip irrigation systems can be managed well, the efficiency of water application was found often better in drip irrigation systems

(Table 1). As seen Table 1, most of the applied water has been used by crops, which are irrigated with drip irrigation systems. Particularly subsurface drip irrigation systems in Middle Anatolian region have been getting the popularity for irrigation of some field crops such as alfalfa due to the great water/energy saving.

Table 1. Water distributions for irrigation systems after irrigation process [13]

Water use (%)	Surface irrigation	Sprinkler irrigation	Drip irrigation
Transpiration	40–70	65–85	85–95
Evaporation	10–25	10–30	5–15
Runoff and deep percolation	15–50	5–15	0–10

Deficit irrigation of certain crops could be another method that is both water-efficient and productive. Gaining a high yield per unit of irrigation water is the primary objective of deficit irrigation. When employing this tactic, crops are either prevented from absorbing adequate water during critical growth stages or their water needs intermittently or drastically lowered during irrigation events [14]. Deficit watering based on plant growth cycles is often possible [15]. Our studies on sugar beets, maize, potatoes and dry bean crops on the Konya plain found that deficit irrigation of up to 25% resulted in no discernible decrease in yield compared to full irrigation [16]. There was an also field research conducted by our team relevant to the water application of different plant growth stages affect on yield and some yield components of drip irrigated sunflower plant in Research Station of Agricultural Faculty of Selçuk University, Turkey for 2016 and 2017 plant growing seasons [17]. For that purpose a total four-irrigation treatments namely irrigation at all plant growth cycles (irrigation at vegetative, flowering and pod filling, VFPF), at only vegetative stage (V), at only flowering stage (F), and at only pods filling stage (PF) were examined. Total applied irrigation water and evapotranspiration, ETc, were between 530 mm (VFPF) and 164 mm (V), and 662 mm (VFPF) and 326 mm (V), respectively. The maximal seed yield was found 4911 hg/ha from VFPF while the lowest one as 1711 kg/ha from PF irrigation treatments (Table 2). As a result, only one irrigation at flowering stage led to 3060 kg/ha seed yield so sunflower plant cannot be exposed to water deficit conditions particularly at flowering cycle.

In other study [18] it was clearly stated that determination of critical stages of crops to the different water deficiency regimes was significance at agricultural water management in water scant regions. This statement is inline with study performed by Acar [17].

Comparable results were found in a study of the impact of full irrigation, FI, 80% of FI, and 60% of FI on the biomass and seed yield of drip-irrigated maize plants grown in sandy loam soil in hot, semiarid climates [19] in northern Togo from November 2017 to April 2018 revealed that FI and 60% of FI treatments produced the highest (2200.4 kg/ha) and the lowest (1068.3 kg/ha) grain yields, respectively. When comparing grain yield at FI and at 60% of FI, there were noticeable changes. It was determined that there were no appreciable changes in yield between FI and 80% of FI, hence 20% deficit irrigation was advised for arid locations like Togo.

Table 2. Seed yield and yield components [17].

Treatments	Seed yield (kg ha^{-1})	Head diameter (cm)	Plant height (cm)	1000-seed weight (g)
VFPF	4911a	15.9a	107.5a	81.3a
V	2448bc	14.7a	107.0a	47.4c
F	3060b	14.3a	86.7b	56.4b
PF	1711c	11.9b	84.0b	35.4d
LSD0.05	1252	2.10	11.85	8.69
CV (%)	20.7	7.4	6.2	7.9

In another trial conducted in the Melkassa area of Ethiopia, furrow-irrigated maize produced 8.4 t/ha at maximum yield when 100% of crop water was applied, while there was no significant difference in production between 100% and 86% of crop water application [20]. In studies performed at Konya province showed that up to 25% deficit irrigation for some crops could be applied for maximal water productivity [16, 21].

Previous results from our research are consistent with those found in other studies conducted in arid and semi-arid areas.

4 Recommendations

Recent years have seen detrimental impacts on the sustainability of both surface and groundwater in most parts of Turkey due to climate change. In the Konya region of Turkey, the intensive agricultural water pumping has led to a drastic decrease in groundwater levels.

Converting water delivery systems to pipe systems could reduce losses when transporting water to irrigation regions. Distributed use of sprinkler and drip irrigation systems, which are highly efficient with water, is strongly encouraged. If we want to boost crop yields, we need to use marginal waters like sewage and rainfall gathering technologies to irrigate more land. Rich visual materials should be used to teach farmers about water management in agriculture. Increases in water fee also increase water production while reducing demand on scarce water supplies [22].

5 Conclusion

In semi-arid Konya plain, withdrawal of groundwater greatly has exceeded rate of recharge. The reasons behind those were lack of surface water bodies, and increments of crop pattern in favor of high water consuming crops such as seed / silage maize. In that case, it is inevitable to face the environmental problems such as some formations of sinkholes associated by over water use from the groundwater reservoir. We need urgent attempts to use water resources sustainable particularly in agriculture. In that regard, we should organize our crop patterns by considering current water resources. The land size of low water consuming crops such as all cereals, sunflower, and legumes should be enlarged. The landowners producing low water consuming crops must be subsidized more. In addition, deficit irrigation should be introduced to farmers by some training activities at farm levels. In that context, we strongly recommend up to 25% deficit irrigation for putting more areas into agro-production in regions where water shortage is remarkable problem. Studies also should focus on developments of new crop cultivars, which are very tolerant to the water stress conditions. Farmers should pay irrigation water fee in volumetric basis. Public should be informed by efficient media about how to improve water productivity particularly in water shortage lands such as Konya province.

Acknowledgements. The data were mainly obtained from the field trials relevant to water-yield relationships of some crops under different water-stress conditions performed by my colleagues from Departments of Farm Buildings & Irrigation, Faculty of Agricultural, University of Selçuk, Turkey. My particular thanks go to all for giving me those data.

References

1. Acar, B., Hamur, Ş: Procedures for successful agricultural water Management in water poor environments under climate changes. Int. J. Agricult. Econ. Dev. **7**(2), 26–32 (2019)
2. Acar, B., Yılmaz, A.M., Kalender, M.A.: Successful agricultural water management for water poor environments like Konya Basin. Turkey. Ann. University Craiova **23**(64), 310–313 (2018)
3. Chartzoulakis, K., Bertaki, M.: Sustainable water management in agriculture under climate change. Agric. Agric. Sci. Procedia **4**, 88–98 (2015)
4. Jehan, S., Iqbal, M., Samreen, T., Liaquat, M., Kanwal, S., Naseem, M.: Effect of deficit irrigation practice on nitrogen mineralization and nitrate nitrogen leaching under semi-arid conditions. J. Water Resour. Prot. **14**, 385–394 (2022)
5. Zhang, S., Wang, X., Zhou, L.: A Review on water-saving agriculture in Europe. J. Water Resour. Prot. **14**, 305–317 (2022)
6. Ramirez, O.A., Ward, F.A., Al-Tabini, R., Philips, R.: Efficient water conservation in agriculture for growing urban water demands in Jordan. Water Policy **13**, 102–124 (2011)
7. Cihan, İ, Acar, B.: Performance of ova water user association in Konya – Turkey. World J. Innov. Res. **1**(2), 25–28 (2016)
8. Sah., R.C.: Groundwater Depletion and Its Impact on Environment in Kathmandu Valley. A Technical Report, 20 ps, August, 2001 (2001)
9. Karimi, V., Karami, E., Keshavarz, M.: Climate change and agriculture: impacts and adaptive responses in Iran. J. Integr. Agric. **17**, 1–15 (2018)
10. Kuwayama, Y., Thompson, A., Bernknopf, R., Zaitchik, B., Vail, P.: Estimating the impact of drought on agriculture using the U.S. drought monitor. Amer. J. Agr. Econ. **101**, 193–210 (2018)
11. Acar, B., et al.: Irrigation in various growth strategies effect on yield and water productivity of drip-irrigated sunflower in semi-arid Konya environment, Turkey. Int. J. Environ. Agric. Res. **6**(12), 91–94 (2020)
12. Acar, B., Direk, M., Yurteri, Y.S., Sayman, O., Muradi, S.M.: Field water management for saving water in water-starved environments such as Konya plain. Türkiye. Green Rep. **3**(8), 48–50 (2022)
13. Grafton, R.Q., et al.: The Paradox of Irrigation Efficiency: Higher Efficiency Rarely Reduces Water Consumption. Policy Forum, Water, May, 2020 (2020)

14. Yavuz, D., Acar, B., Yavuz, N., Çiftçi, N.: Aspects for agricultural water management in water stress conditions: Case Study of Konya Plain, Turkey. Int. J. Agric. Econ. Dev. **6**(2), 7–17 (2018)
15. Zhang, H., Han, M., Comas, L.H.: Response of maize yield components to growth stage-based deficit irrigation. Agron J. **3**(6), 3244–3252 (2019)
16. Acar, B., Topak, R., Yavuz, D., Kalender, M.A.: Is drip irrigation technique sustainable solution in agriculture for semi-arid regions? A case study of Middle Anatolian Region, Turkey. Int. J. Agric. Econ. Dev. **2**(2), 1–8 (2014)
17. Yavuz, D., Acar, B., Yavuz, N., Çiiftçi, N.: Irrigation in various growth stages effect on yield and water productivity of drip- irrigated sunflower in semi-arid Konya environment, Turkey. Int. J. Agric. Econ. Dev. **6**(2), 7–17 (2018)
18. Acar, B.: Deficit Irrigation Effect on Water Use Efficiency of Crops in Arid and Semi-Arid Regions. Int. J. Agric. Econ. Dev. **7**(2), 18–25 (2019)
19. Gadedjisso-Tossou, A., Avellan, T., Shütze, N.: Impact of irrigation strategies on maize (Zea mays L.) production in the savannah region of northern Togo (West Africa). Water SA, **46**(1), 141–152 (2020)
20. Seid, M.M., Narayanan, K.: Effect of deficit irrigation on maize under conventional, fixed and alternate furrow irrigation systems at Melkassa, Ethiopia. Int. J. Eng. Res. Technol. **4**(11), 119–126 (2015)
21. Jalal Jalal, O.A., Acar, B.: Water use in sprinkler-irrigated carrot plant in semi-arid Konya-Kasınhanı province of Turkey. World J. Innov. Res. **4**(2), 1–4 (2018)
22. OECD: Sustainable Management of Water Resources in Agriculture. Report, 118 ps (2010)

Check for updates

Some Considerations on the Application of Ocean Wave Energy for Water Pumping in Near Shore Areas in Mozambique Channel

Alberto Filimão Sitoe[1,2], António Mubango Hoguane[3,4], and Soufiane Haddout[5](✉)

[1] Faculty of Natural and Exact Science, Save University (UniSave), Chongoene, Mozambique
[2] Department of Physics, Faculty of Science, Eduardo Mondlane University (UEM), Maputo, Mozambique
[3] Centre for Marine Research and Technology (CePTMar), Eduardo Mondlane University, P.O.Box 128, Quelimane, Mozambique
[4] Mozambique Oceanographic Institute (InOM), Maputo, Mozambique
[5] Department of Physics, Faculty of Science, Ibn Tofail University, B.P. 133 Kenitra, Morocco
soufian.haddout@gmail.com

Abstract. Previous studies have indicated that off Mozambique coast is characterized by low to moderate and high variability ocean wave climate, with considerable time during the year with wave energy below the threshold for electricity production. This paper discusses the potential for applications of the available ocean wave energy off Mozambique coast, for water pumping and for saltwater desalination. Wave energy absorption efficient was set to 25%. The saltwater pumping rate varies from 184.5 ± 130.5 L per hour per a meter of wave width (L/hr/m), and the freshwater production rate in a wave powered reverse osmosis desalination plant varied from 338.3 ± 239.3–1,377.8 ± 1,196.3 L/hr/m, with high values observed in the southern part of the channel, and low values were observed in the northern part of the Channel. Saltwater pumped into coastal reservoirs may be used for inland aquaculture and mini-hydroelectric power stations and freshwater produced may be used for drinking, sanitation and irrigation, all of which with potential to boost the livelihood, income generation and economy of the coastal areas. Harvesting of wave energy of up to 100 m wave width would yield 15 tons of fish, 90,000 mussels per year and 15 kW electricity production, in southern Channel and providing portable water for 13,800–45,000 living in coastal areas adjacent to Mozambique Channel.

Keywords: Wave energy · Water pumping · Electricity · Desalination · Fish farming

1 Introduction

Wave energy has a great potential for boosting the livelihood, income generation and economy in coastal areas [1, 2]. However, most effort for harvesting wave energy have focused on direct application for generating electricity [3–5], despite other potential applications.

S. Haddout et al. (Eds.): CCORE 2022, SPEES, pp. 171–178, 2023.
https://doi.org/10.1007/978-3-031-26967-7_14

Previous studies have indicated that Mozambique Channel has a moderate wave energy potential, with more energetic waves observed in the southern part of the Channel (1.5–2 ± 0.8 m, 8–20 ± 1.7 s and 10–23 ± 6.3 kW m^{-1}, on average, for significant wave height, wave period and wave energy flux, respectively), followed by northern part of the Channel (1.2–1.4 ± 0.5 m, 6–8 ± 1.7 s, 6–7 ± 3.9 kW m^{-1}, on average), with the middle part displaying a lower energy (6 ± 3.2 kW m^{-1}, on average) [6]. However, the wave climate was found to be highly variable (≥80%), with the minimum threshold of significant wave height (1.5 m) and wave energy flux (10 kW m^{-1}) for electricity production attained and exceeded with a probability of 61–63%, which makes the application of wave energy for direct production of electricity unreliable [7]. Notwithstanding the fact that Mozambique Channel is fringed mostly by low income countries which cannot afford high technology and costly devices. Hence, the study suggested the exploration of the applications of wave energy resources for desalination and water pumping, with potential to enhancing the livelihood of many people living in coastal areas.

Subsequently, the present study discusses the possibility of direct application of wave energy for water pumping on coastal areas. The system is well described by [8], it consists of a moored floating structure, fixed at the sea bed, connected to a piston pump. The pump, move vertically under the action of the waves, pushes water into coastal storage reservoirs located few meters (~10 m) above sea level. Some prototype systems may push water up to 14 m above the Mean Sea Level at the rate of 3 L per minute, or at the rate of 1.5 L per minute up to 25 m height [9]. Therefore, a set of devices may pump huge amounts of water per day. The system employs low-tech and low-cost devices, and so, easy to maintain [8, 9], suiting better the rural conditions of low income countries. The stored water can then be used for a variety of coastal applications such as coastal fish farming, coastal swimming pools, in addition, it can be used for hydro power generation [9]. The storage of water in uplifted reservoirs is a low-cost mean of energy storage compared with the use of batteries, for the case of electricity production, for instance.

Furthermore, wave powered pumps can be used to drive saltwater through reverse osmotic desalination plant and produce freshwater [10, 11]. This system result in reduced costs for desalination, against traditional pumps that use electrical or diesel pumps. The freshwater produced can be used for drinking, sanitation and irrigation, and so, addressing the water scarcity [10] in coastal areas, were estuaries are affected by salt intrusion [12, 13] caused by climate change impacts.

2 Methodology

2.1 Description of the Study Site

The present study was conducted in Mozambique Channel, an open channel, 400 km to 950 km width and average depth of 3,292 m, oriented north-southwards from about 10°S to 26°S, in the Western Indian Ocean (WIO) sub-region (Fig. 1). The wave climatology is dominated by swells generated in the southern ocean and propagating northwards through the channel and swell waves generated by the northern winds, penetrate the channel from the northern boundary and propagate southwards [14].The wave climatology and wave energy near the coast in Mozambique Channel was described by [6].

2.2 Data Set

The present study used average wave energy flux, over a 2.5 years period (from 29th June 2019 to 31st December 2021), estimated from weekly average of daily significant wave height and wave period, measured from satellite altimeters, provided by marine analyst (http://www.marine-analyst.eu), as estimated by [6], in 10 selected sites in Mozambique Channel, whose coordinates are given in Table 1 (see Fig. 1).

2.3 Technical Viability of Wave Energy for Water Pumping

Technical viability of wave energy for salt water pumping was estimated considering a moderate water pumping rate, up to 10 m height, of 180 L per hour and per kilowatt of wave energy, as indicated by [9], and freshwater production rate in the reverse osmosis desalination plant of 330 L per hour and per kilowatt, as indicated by [15]. The overall wave energy power absorption efficiency of the conversion systems was set to an average value of 25% based on estimates by [16]. Hence, the gross pumping rate of saltwater to 10 m height, powered by wave energy, is given by:

$$Q^{\bullet}_{10m}[L/hr] = 0.25 \times 180[L/hr/kW] \times P_W[kW/m] \quad (1)$$

where $Q^{\bullet}_{10m}$ is the pumping rate of saltwater up to 10 m height in litre per hour, per a meter of wave width, and P_W is the bulk wave energy flux, given in kilowatts per meter of wave width. The freshwater production rate in a reverse osmosis plant driven by wave energy is given by:

$$\overset{\bullet}{Q}[L/hr] = 0.25 \times 330[L/hr/kW] \times P_W[kW/m] \quad (2)$$

where $\overset{\bullet}{Q}$ is the freshwater production rate in an osmosis desalination plant given in litre per hour, per a meter of wave width.

3 Results

Table 2 presents the estimate of saltwater pumping rate of a wave powered pump and freshwater production rate in wave powered reverse osmosis desalination plant as a function of available wave power per a meter of wave width.

The saltwater pumping rate, estimated by Eq. 1, varied from 184.5 ± 130.5 L per hour per a meter of wave width (L/hr/m) in Moroni to 1,021.5 ± 774.9 L/hr/m in Toliara. High pumping rates (603.0 ± 1,044.0–1,021.5 ± 774.9 L/hr/m) were observed in southern part of the Mozambique Channel in Inhambane, Morondava and Toliara. Low pumping rates (184.5 ± 130.5–315.0 ± 342.0 L/hr/m) were observed in the northern part of the Channel, in Nosy Be, Mayotte, Maroni and Pemba. The middle part of the Channel displayed mid values (472.5 ± 454.5–508.5 ± 418.5 L/hr/m).

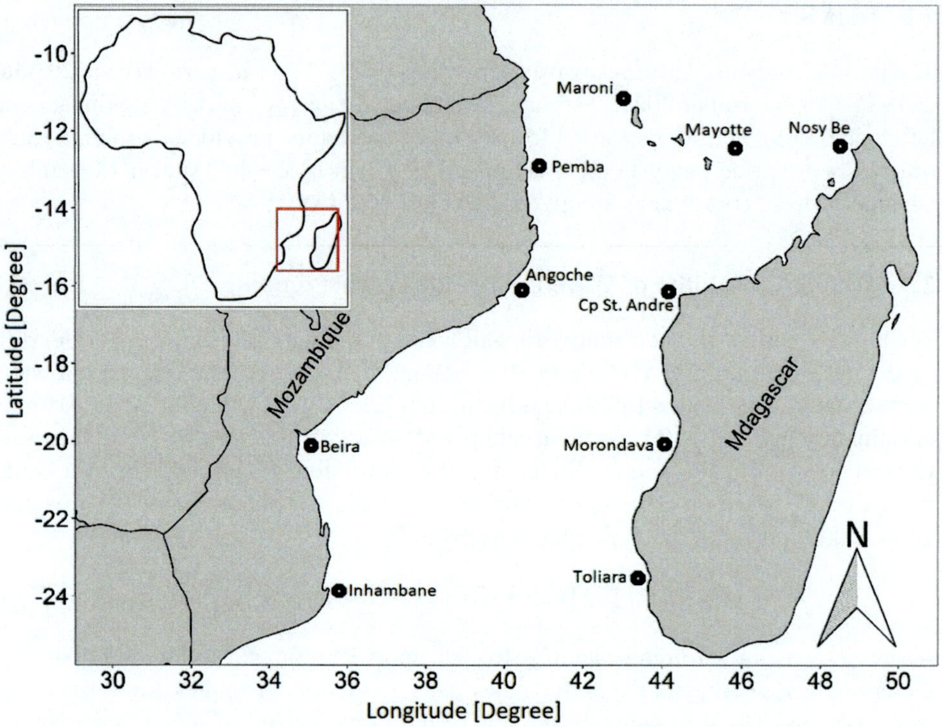

Fig. 1. Location of the study area (after [6]).

Table 1. Coordinates of the quadrats where the wave data were retrieved (after [6]).

Region	Local	Latitude °S		Longitude °E	
		Initial	Final	Initial	Final
Western side of the Channel	Moroni	11.37	12.01	42.85	43.24
	Pemba	10.92	13.28	40.25	42.45
	Angoche	16.09	16.09	39.86	40.25
	Beira	18.96	20.92	34.61	37.24
	Inhambane	21.23	23.50	35.11	37.57
Eastern side of the Channel	Mayotte	11.32	11.97	43.45	43.85
	Nosy Be	12.96	13.61	47.80	48.19
	Cape St. Andre	15.80	16.44	44.42	44.81
	Morondava	19.95	20.60	43.87	44.26
	Toliara	23.33	23.66	43.51	43.65

The freshwater production rate in a wave powered reverse osmosis desalination plant, estimated by Eq. 2, varied from 338.3 ± 239.3–1,377.8 ± 1,196.3 L/hr/m. High freshwater production rates (1,105.5 ± 1,196.3–1,377.8 ± 1,196.3 L/hr/m) were observed in southern part of the Channel, in Morondava and in Inhamabne. Low values of freshwater production rates (338.3 ± 239.3–577.5 ± 627.0 L/hr/m) were observed in the northern part of the Channel, in Nosy Be, Mayotte, Moroni and Pemba. The freshwater production rates in the middle of the Channel varied in the range 866.3 ± 833.3–932.3 ± 767.3 L/hr/m) in Angoche and Beira.

Table 2. Estimate of wave energy flux, saltwater pumping rate and freshwater production rates near the coast of the Mozambique Channel.

Region	Site	Wave energy flux		Wave energy absorption efficiency [%]	Saltwater pumping rate		Freshwater production rate	
		Average [kW/m]	Standard deviation [kW/m]		Average [L/hr/m]	Standard deviation[L/hr/m]	Average [L/hr/m]	Standard deviation[L/hr/m]
Western side	Moroni	7.0	5.6	25	315.0	252.0	577.5	462.0
	Pemba	7.0	7.6	25	315.0	342.0	577.5	627.0
	Angoche	10.5	10.1	25	472.5	454.5	866.3	833.3
	Beira	11.3	9.3	25	508.5	418.5	932.3	767.3
	Inhambane	16.7	14.5	25	751.5	652.5	1377.8	1196.3
Eastern side	Mayotte	6.1	5.2	25	274.5	234.0	503.3	429.0
	Nosy Be	4.1	2.9	25	184.5	130.5	338.3	239.3
	Cpe St. Andre	4.5	3.5	25	202.5	157.5	371.3	288.8
	Morondava	13.4	23.2	25	603.0	1044.0	1105.5	1914.0
	Toliara	22.7	17.2	25	1021.5	774.0	1872.8	1419.0

4 Discussion

4.1 The Use of Pumped Saltwater for Marine Culture

Marine area is large and marine water are abundant, and so, offering huge potentials for development of marine culture compared to inland water in the region, yet, the cost of offshore marine culture is enormous, which constrains the poor local community to harness these potentials. Hence, land-based marine culture, with pumped water, with or without recirculation was suggested as a viable alternative for the production of high quality seafood in low income countries [17].

The *water requirement* for sustainable production in aquaculture ponds, considering the evaporation loss and see-page loss and water exchanges requirements, was estimated at 10.3 m^3/Kg of fish production [18]. Accordingly, taking into consideration the potential for saltwater pumping rates as estimated in Table 2, the northern part of the Channel with lowest potential, a device that pumps 315 L of seawater per hour, and considering

6 months of production cycle would enable a sustainable fish production of the order of 150 kg per wave energy of a meter wave width. The southern part of the Channel, with the highest wave energy potential, the production rates would increase to about 0.5 tons per wave energy of a meter wave width. Hence, wave energy of 100 m wave width would yield about 15 tons of fish per year.

The mussel production is more demanding in terms of water quality and water flow. Papadimitriou *et al.* (2021) [19] estimated a flow rate of about 5–10 cm/s for a sustainable mussel culture activity. Considering stocking density of 4 mussels/m^3 as the optimal breeding density as suggested by [20], the southern part of the Channel (Inhambane and Toliara) offers the ideal place for promoting coastal based mussel culture; even though, they would require an scaling up of the devise to a series 15–25 pumps operating at wave energy of a meter wave width, which would yield 4 mussels per year. Scaling this up to a hector, which would yield 90 thousand mussels, it would require 2.25 m^3 per day, and it would require a system that would pump saltwater at the rate of 26 m/s, or a harvesting of wave energy of up to 100 m wave width in southern Channel.

4.2 The Use of Pumped Saltwater for Mini-Hydropower Operation

Pumped saltwater to an elevated reservoir would enable the production of electricity through a mini-hydropower. The potential power can be calculated as follows [21, 22]:

$$P = g * Q * H * f_{eff} \tag{3}$$

where, Power in kW (P); Flow rate in m^3/s (Q); Head in m (H); Gravity constant = 9.81 m/s^2 (g); Efficiency factor (f_{eff}) = > 0.4–0.7, here set to a minimum value. Hence, considering pumping water to a head of 10 m, and flow of 300 L/s (=0.3 m^3/s) will have a potential power of 15 kW electricity; which would require harvesting wave energy of 100 m wave width in Southern Mozambique channel.

4.3 Freshwater Production

Human water consumption, in developing countries, was estimated at 92 L/capita/day [23]. Subsequently, considering the estimated freshwater production through reverse osmosis plant operated by wave power, presented in Table 2, it can be seen that a wave energy of one meter wave width would provide water for about 100 to 450 people in the least and most wave energetic side s of the Channel, respectively. Considering wave energy potential of 100 m wave width it could provide water for 13,800 to nearly 45,000 people living in coastal areas.

5 Concluding Remarks

5.1 The Use of Pumped Saltwater for Marine Culture

This discusses the use ocean wave energy to pumping water for variety of applications, in coastal areas around Mozambique Channel, based on the estimate of wave energy potential by [6]. Salt water pumped to a reservoir up to 10 m above the mean sea level and into reverse osmosis desalination plant was investigated. The study suggested the use of salt water pumped into reservoir in marine culture and in mini-hydropower production. The found out that harvesting wave energy of 100 m wave width would yield 15 tons of fish and 90,000 mussels per year and 15 kW of electricity. With regard to freshwater production in a wave powered desalination plant, the study estimated that wave energy potential of 100 m wave width would be sufficient to pump water that would benefit 13,800 to about 45,000 people in coastal areas. Hence, the study concludes that wave energy, yet, exploited with low-tech and low-cost technology may contribute significantly to improved livelihood, family income and economy of the people living in coastal areas.

Acknowledgement. The research was part of PhD studies on renewable energy, financially supported by the Eduardo Mondlane University post-graduate research grant, under SIDA-SAREC Project No 2.1.6.

References

1. Lavidas, G.: Energy and socio-economic benefits from the development of wave energy in Greece. Renewable Energy **132**, 1290–1300 (2019). https://doi.org/10.1016/j.renene.2018.09.007
2. Mouakkir, L, El hou, M., Mordane, C.M.: Wave energy potential analysis in the casablanca-mohammedia coastal area (Morocco). J. Marine Sci. Appl. **21**, 92–101 (2022). https://doi.org/10.1007/s11804-022-00261-2
3. Mattiazzo, G.: State of the artand perspectives of wave energy in the Mediterranean sea: backstage of ISWEC. Front. Energy Res. **2019**(7), 114 (2019). https://doi.org/10.3389/fenrg.2019.00114
4. Aderinto, T., Li, H.: Ocean wave energy converters: status and challenges. Energies **11**(5), 1250 (2018). https://doi.org/10.3390/en11051250
5. Babu1, N., Balaji, K.K., Nishal, S.: Wave energy for desalination plants –a review. Int. J. Eng. Res. Technol. (IJERT) **5**(7) (2017)
6. Sitoe, A.F., Hoguane, A.M., Haddout, S.: Preliminary assessment of near-shorewave energy potential in the Mozambique Channel. J. Water Clim. Change **00**, 1 (2022). https://doi.org/10.2166/wcc.2022.475
7. Rusu, E., Soares, C.G.: Wave energy pattern around the Madeira Islands. Energy **45**(1), 771–785 (2012)
8. Barbarelli, S., Amelio, M., Florio, G., Scornaienchi, N.M.: Study of a hydraulic system converting energy from sea waves near the coast. MATEC Web of Conferences **240**, 01004 (2018). ICCHMT 2018: https://doi.org/10.1051/matecconf/201824001004
9. Lee, M.D., Feng, E.L.K., Lee, P.S.: Small scale low height wave energy seawater pump for achieving environmental and economic sustainability. Univ. J. Mech. Eng. **8**(1), 14–20 (2020). https://doi.org/10.13189/ujme.2020.080102
10. Leijon, J., Boström, C.: Freshwater production from the motion of ocean waves – a review. Desalination **435**, 161–171 (2018). https://doi.org/10.1016/j.desal.2017.10.049
11. Davies, P.A.: Wave-powered desalination: resource assessment and review of technology. Desalination **186**, 97–109 (2005). https://doi.org/10.1016/j.desal.2005.03.093
12. Little, S., Lewis, J.P., Pietkiewicz, H.: Defining estuarine squeeze: The loss of upper estuarine transitional zones against in-channel barriers through saline intrusion. Estuar. Coast. Shelf Sci. **278**, 108107 (2022). https://doi.org/10.1016/j.ecss.2022.108107

13. Pereira, H., Sousa, M.C., Vieira, L.R., Morgado, F., Dias, J.M.: Modelling salt intrusion and estuarine plumes under climate change scenarios in two transitional ecosystems from the NW Atlantic coast. J. Marine Sci. Eng. **10**(2), 262 (2022). https://doi.org/10.3390/jmse10020262
14. Semedo, A., Sušelj, K., Rutgersson, A., Sterl, A.: A Global View on the wind sea and swell climate and variability from ERA-40. J. Clim. **24**(5), 1461–1479 (2011). https://doi.org/10.1175/2010JCLI3718.1
15. Viola, A., Franzitta, V., Trapanese, M., Curto, D.: Nexus water & energy: a case study of wave energy converters (WECs) to desalination applications in Sicily. Int. J. Heat Technol. **34**, S379–S386 (2016). https://doi.org/10.18280/ijht.34S227
16. Aderinto, T., Li, H.: Review on power performance and efficiency of wave. Energy Conv. Energies **12**, 4329 (2019). https://doi.org/10.3390/en12224329
17. Soto, D., Wurmann, C.: Offshore aquaculture: a needed new frontier for farmed fish at Sea. In: The Future of Ocean Governance and Capacity Development, pp. 379–384 (2018). https://doi.org/10.1163/9789004380271_064
18. Sharma, K.K., Mohapatra, B.C., Das, P.C., Sarkar, B., Chand, S.: Water budgets for freshwater aquaculture ponds with reference to effluent volume. Agric. Sci. **4**(8), 353–359 (2012). https://doi.org/10.4236/as.2013.48051
19. Papadimitriou, C.A., Savvidis, Y.G., Galinou-Mitsoudi, S., Moriki, A.: The role of hydrodynamics on the sustainable mussels' culture activity. The Case of the Chalastra Basin (NW Gulf of Thessaloniki). Hydrology **8**(3)105. https://doi.org/10.3390/hydrology8030105
20. Geng, B., Li, Y., Liu, X., Ye, J., Guo, W.: Effective treatment of aquaculture wastewater with mussel/microalgae/bacteria complex ecosystem: a pilot study. Sci. Rep. **12**(1), 2263 (2022). https://doi.org/10.1038/s41598-021-04499-8
21. Micro-hydro Design Manual, A Harvey & A Brown. ITDG Publishing (1992)
22. Fraenkel, P., Paish, O., Bokalders, V.: Micro-hydro power: A guide for development workers. A Harvey & A Brown, ITDG Publishing, IT Power, Stockholm Environment Institute (1991)
23. Crouch, M.L., Jacobs, H.E., Speight, V.L.: Defining domestic water consumption based on personal water use activities. J. Water Supply Res. Technol. AQUA **70**(7), 1002–1011 (2021). https://doi.org/10.2166/aqua.2021.056

Predicting the Spatio-Temporal Evolution of DO and COD in the Bouregreg Estuary (Morocco): First Results

Soufiane Haddout[1](✉), K. L. Priya[2], Joan Cecilia C. Casila[3], Mary Ann Q. Franco[4], and António Mubango Hoguane[5,6]

[1] Department of Physics, Faculty of Science, Ibn Tofail University, B.P. 133 Kenitra, Morocco
soufian.haddout@gmail.com
[2] Department of Civil Engineering, TKM College of Engineering, Kollam 691005, India
[3] Land and Water Resources Division, IABE, CEAT, University of the Philippines Los Baños, 4031 College, Laguna, Philippines
[4] Aquatera Asia Pte Ltd, Singapore, Singapore
[5] Centre for Marine Research and Technology (CePTMar), Eduardo Mondlane University, P.O.Box 128, Quelimane, Mozambique
[6] Mozambique Oceanographic Institute (InOM), Maputo, Mozambique

Abstract. In Morocco, the Bouregreg estuary is located between Rabat and Salé at 34°2′9″N and 6°50′7″W, from its origins in the Middle Atlas Mountains to the Atlantic Ocean, it travels around 240 km. It is a part of a watershed with an area 9800 $km^2$and an elevation range of 0 to 1627 m. This estuary is dominated by seawater, with minimal freshwater inflow for most of the year. Agricultural and industrial areas are located near the estuary mouth. Some urban sewage from Rabat and Sale is discharged directly into the estuary. According to our knowledge, there is no research on the water quality modeling in the Bouregreg Estuary. For the first time, dissolved oxygen (DO) and chemical oxygen demand (COD) are calculated in this study using an adopted 1-D numerical model (AQUASIM) based on reaeration rates and measured concentrations of DO and COD. Various statistical parameters were used to assess the accuracy of the AQUASIM model. The findings indicate that autumn has higher concentrations of DO along the estuary (from 3.18 mg/l at 7 km to 8.5 mg/l) than summer (from 2.97 mg/l at 7 km to 8 mg/l at the mouth). Additionally, from 39 mg O2/l at the mouth to 236 mg O2/l at 7 km, the average COD values increased. The present work will help to assess the variability and suitability of aquatic habitats, thereby providing enhanced information about the Bouregreg estuary ecosystem and a deeper understanding of its dynamics.

Keywords: Numerical model · Bouregreg estuary · Dissolved oxygen · Chemical oxygen demand

1 Introduction

Estuaries are classified and defined based on their use, but a wide variety of definitions and classifications exist [1, 2]. Typically, tides, waves, rivers, sediment types, sediment

S. Haddout et al. (Eds.): CCORE 2022, SPEES, pp. 179–189, 2023.
https://doi.org/10.1007/978-3-031-26967-7_15

supply, vegetation, geology, etc. are used to define estuaries [3]. The welfare of aquatic ecosystems, the mixing of river and ocean water, and freshwater supplies are all significantly impacted by estuaries [4]. As a result, estuarine system functioning is a crucial area of study. Dissolved oxygen (DO) or chemical oxygen demand (COD), one of the important water quality indicators, is a crucial indicator of the habitat suitability for aquatic biota and can reflect the health of a river water ecosystem. DO stands for the oxygen dissolved through diffusion from the surrounding air, aeration of water from waterfalls and rapids, and as a byproduct of photosynthesis (in the presence of light and chlorophyll). Most aquatic life requires DO for respiration. In addition, DO forms compounds like carbonate, sulphate, nitrate, and phosphate with other essential elements like carbon, sulfur, nitrogen, and phosphorus that, in the absence of oxygen in water bodies, could have been toxic. These molecules help ensure the survival of aquatic species. Therefore, sufficient oxygen levels are needed to support aerobic life forms during natural stream cleaning processes. Aquatic life is put under physiological stress if the DO content in the water drops below 5.0 mg/l. For a few hours, oxygen concentrations below 1–2 mg/l can have an impact on aquatic species' growth and survivability. Therefore, DO is a necessary component for all types of aquatic life [5]. Most analyses of the Bouregreg Estuary (Morocco) concentrate on: pollution assessment [6, 7], salt-intrusion [8, 9], observation of water quality variables [10], sediment/ heavy metal studies [11–13]. As the Bouregreg estuary is important to nearby communities, studying DO/COD modeling isn't yet being conducted. It is crucial to have a detailed understanding of DO/COD processes in estuaries in order to support policy and management decisions.

Models have been used in numerous studies of DO/COD. Numerical models, particularly two-dimensional (2-D) and three-dimensional (3-D) models, are more common today because of their capacity to provide greater spatial and temporal detail [14, 15]. Numerical models, or 1-D mathematical models, are the best tool for quick scan operations during the planning stage of a project or for instructional purposes. Furthermore, it is methodologically sound to begin with a simple explanation of the phenomenon, consider its limitations, and then move on to consider further difficulties. We initially used a one-dimensional hydraulic water quality model to study the hydraulic model for system dynamics (AQUASIM). The DO and COD evolution in the Bouregreg estuary was examined. Data measurements for the selected model were taken between 2021 and 2022.

2 Numerical Model

2.1 Hydraulic Model

Using cross-sectional profiles collected inside the sample reach, a constant slope, and the cross-sectional profiles recorded within the study reach, an effective friction coefficient was computed. Diffusive wave approximations were used to compute river hydraulics in accordance with de Saint Venant's equations. In AQUASIM [16], river hydraulics is described using two main approximations to these equations. In river hydraulics, dissolved or suspended substances are transported by means of advection-diffusion equations. As a result, we have the following two differential equations. Water flows through the compartment according to the first equation:

$$\frac{\partial A}{\partial t} + \frac{\partial Q}{\partial x} = q \tag{1}$$

A is determined by the spatial-gradient of the discharge Q, and by the lateral-in flow q. The second equation describes how substances moved by water flow behave:

$$\frac{\partial AC}{\partial t} = -\frac{\partial}{\partial x}(QC) + \frac{\partial}{\partial x}\left(AE\frac{\partial C}{\partial x}\right) + r + S_{qn} \tag{2}$$

Advection of the water flow affects the concentration (first term), longitudinal dispersion (second term), transformations (third term), and lateral inflows and outflows (fourth term). Equation (1) and partial differential Eq. (2) require boundary conditions in order to make the solution unique. In Eq. (2), the boundary conditions are provided by the continuity of the substance mass flows as they enter the river section, as well as a transmission boundary condition at the end of the compartment:

$$Q(x_s)C - AE\frac{\partial C}{\partial x} = I_{in,C} \tag{3}$$

$$\frac{\partial^2 C}{\partial x^2} = 0 \tag{4}$$

where $I_{in,C}$ is the total mass input of substance described by the concentration C per unit time. The 2nd of these boundary conditions, Eq. (4), is omitted for dispersion-free transport. According to Fischer et al. 1979 [27], the longitudinal dispersion-coefficient (E) is calculated as follows in the second part of Eq. (2):

$$E = c_f\frac{w^2v^2}{u^*d} \tag{5}$$

where d is the average depth, w is the surface-width of the river, C_f is a nondimensional dispersion coefficient and u^* is the friction-velocity. The friction-slope is determined using the empirical formula (6):

$$S_f = \frac{1}{K_{st}^2}\frac{1}{R^{4/3}}v^2 \tag{6}$$

To calculate the friction slope a Manning-Strickler formula is used. Where R is the hydraulic radius of the river, K_{St} is the friction coefficient according to Strickler and v is the average cross-sectional flow velocity.

2.2 Oxygen Series Modeling

Seawater predominates in the Bouregreg estuary for the majority of the year, with little freshwater inflow. There are agricultural and industrial districts close to the mouth. High daily and seasonal temporal dynamics are a defining characteristic of the oxygen balance of the Bouregreg estuary. It can be represented using an AQUASIM model by applying Eq. (2):

$$r = K_2(C_{sat} - C_{O_2}) + \frac{PI}{d} - \frac{R}{d} - K_{deg}\frac{C_{O_2}}{K_{O_2} + C_{O_2}}COD \quad (7)$$

where K_2 is the reaeration rate constant, C_{O2} is the oxygen concentration, d is the mean river depth, r is the net oxygen production rate, C_{sat} is the oxygen saturation concentration, I is the light intensity, P and R are production and respiration parameters respectively, K_{O2} is the half saturation concentration with respect to oxygen, K_{deg} is the degradation rate constant and COD is the concentration of substrates.

3 Materials and Methods

In Morocco, the Bouregreg estuary is located between Rabat and Salé at 34°2′9″N and 6°50′7″W. Sidi Mohammed Ben Abdellah Dam is located upstream; Rabat and Sale beaches are located downstream. It has the appearance of an arm coming from the water, measures 23 km in length and 150 m in average width, and extends perpendicular to the 6th meridian and to the level of the 34th parallel. There are six sample stations

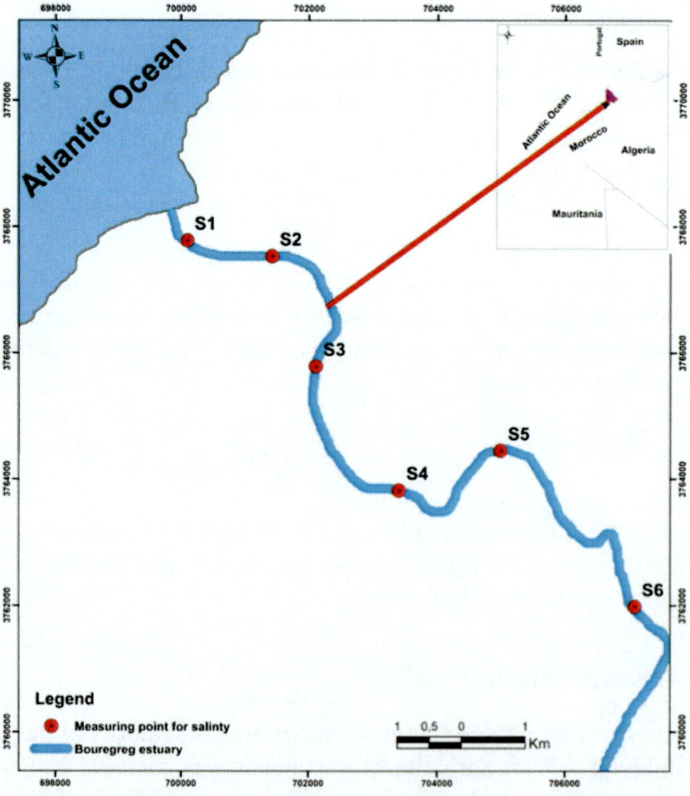

Fig. 1. Map showing Bouregreg estuary and location of measuring stations.

located from upstream to downstream (See Fig. 1). Keeping pace with the tidal wave, the measurements started at the mouth and moved upstream. Study of the water quality of Bouregreg estuary was conducted from Oct-2021 to Sep-2022. The estuary receives some urban sewage directly from Rabat and Sale. Further, two physicochemical parameters were monitored in this estuary in order to evaluate its physicochemical quality: dissolved oxygen (DO) and chemical oxygen demand (COD). The processed water samples were taken to the study desk at Casablanca (Morocco) for analysis.

Using cross-sectional profiles that were gathered over the whole length of the river, a constant slope, and an effective coefficient through each segment of the river, the study of the Bouregreg estuary was modeled. The simulation and data analysis package in AQUASIM [16] was used for all simulations. The website for this program, http://www.aquasim.eawag.ch, has more details.

4 Results and Discussion

One-dimensional modeling was utilized to simulate the evolution of DO and COD along the Bouregreg estuary, which was reasonable given how far the river stretch. The mathematical model AQUASIM was employed in this investigation. The hydraulic regime and water quality have been simulated using AQUASIM model (e.g. [18, 19]). The hydraulic

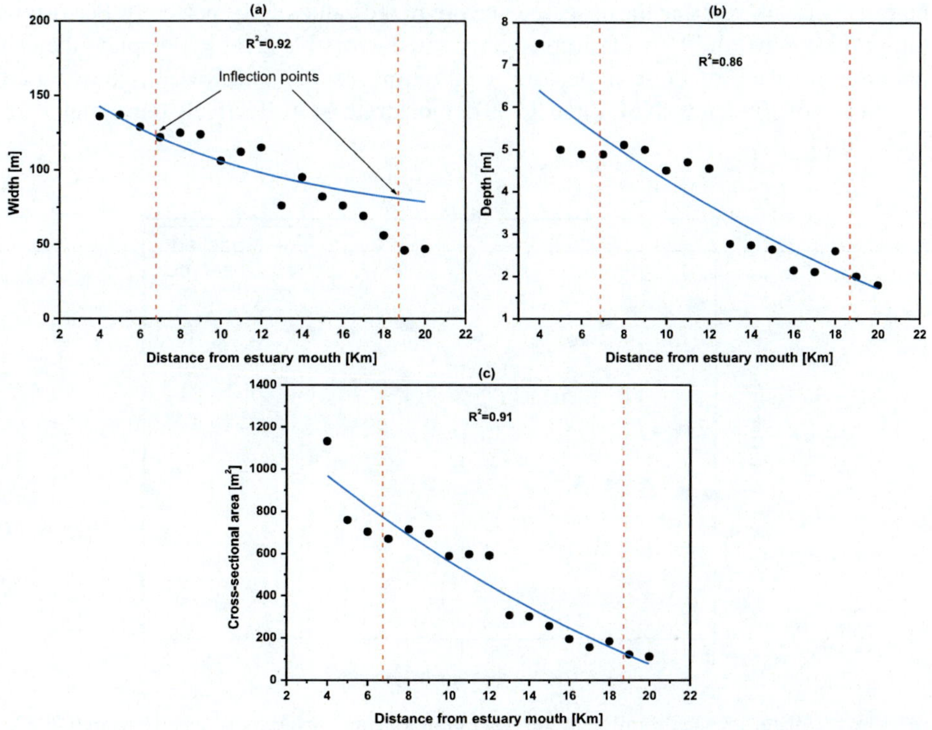

Fig. 2. Longitudinal evolution of a) river width, b) cross-sectional area, and c) depth properties of the Bouregreg estuary

regime of the Bouregreg Estuary, which in turn heavily depends on the river morphology, affects the evolution of DO and COD. For this purpose, the hydraulic regime in AQUASIM was initially explored and analyzed. AQUASIM model developed version was used for all simulations [16]. The results of the hydraulic model were applied to the evolution of DO and COD. Cross-sectional profiles along the whole length of the estuary, a constant slope, and an effective coefficient through each range were used in the modeling of the survey of the Bouregreg estuary. Following a thorough field examination, Fig. 2 depicts the longitudinal evolution of the measured river width, cross-sectional area, and depth.

The Mannings roughness flow serves as the calibration parameter. With an average value 0.068 $m^{1/3}s^{-1}$, the initial Mannings roughness values range along the estuary under study. The Manning coefficient was calibrated along the reach to the same extent since it was thought that the sources of inaccuracy in its evaluation were the same for all grids. The water level data at station S2 is used for calibration and validation. The calibration will take place between March 1 and March 10, 2022. The results of the calibration are shown in Fig. 3. This graph shows that the modeled and observed water levels at station S2 closely match each other. Additionally, the correlation coefficient, the Nash-Sutcliffe Modeling Efficiency Index (NSC), and the root-mean squared error (RMSE) were employed to assess a model's effectiveness (R^2). [20, 21] contains the equations required to calculate these indicators. The RMSE indicates complete agreement between predicted and observed values when it equals 0. The NSC ranges from - ∞ to 1. When this number is 1, it means that the observed and predicted values agree perfectly. Performance ratings between 0 and 1 are often considered satisfactory [21]. Table 1 displays the values obtained for each of these indicators' calibration results. The RMSE, Nash-Sutcliffe coefficient of efficiency (NSC), and R^2 all demonstrate how effectively the model predicts water level.

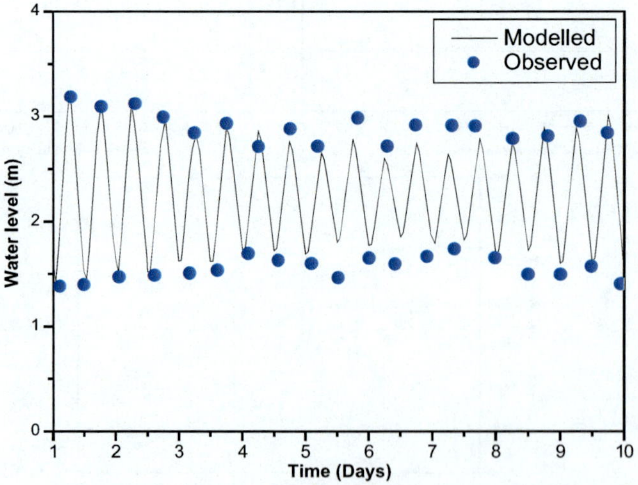

Fig. 3. Water level calibration at the S2 station: during 1 March 2022 to 10 March 2022.

Table 1. Statistical parameters of the hydraulic model

Statistical parameters	RMSE (m)	NSC	R^2
Calibration	0.078	0.87	0.81

The downstream-upstream limits may be the fundamental water quality modeling capability of AQUASIM to forecast the mean observed concentration. Based on empirical estimations [22, 23] that took into account the water levels during the study periods, estimates for physical reaeration were made. The K_2 values using empirical formulas ranged from 17.0 to 29.0 d^{-1}. For model calculations, $K_2 = 24d^{-1}$ was used as the value. [25] provides the formula for the oxygen saturation concentration. From constantly observed oxygen time series, production (P) and respiration (R) rate parameters were calculated based on these boundary conditions (Table 2). A range of 0.5 to 3.0 d^{-1} was suggested as an approximation for K_{deg} in Thomann and Mueller, 1987 [24].

Table 2. Parameter estimates from oxygen time series

Parameters	Value
$P(g/(Wd))$	0.7823
$R(g/(m^2d))$	23.102

The summary of the DO and COD water quality model parameters (according to the observed data along the estuary axis and time series) are shown in Figs. 4 and 5. In order to demonstrate good agreement between the simulated results of the model output and the data along the estuary axis, the performance of the modeling tools was assessed using several statistical methodologies (Fig. 4). Time series observations from the DO and COD were used for validation (Fig. 5). The Bouregreg Estuary's seasonal evolution of dissolved oxygen shows that autumn concentrations (from 3.18 mg/l at 7 km to 8.5 mg/l at the estuary) are greater than those of summer (from 2.97 mg/l at 7 km to 8 mg/l at the mouth). Whereas heavy precipitation during the winter season boosts oxygen exchange with the atmosphere and makes it easier for air to enter the estuary's waters, which results in a dilution of the pollution load. As a result, the Bouregreg Estuary's curve representing geographic variation in dissolved oxygen has the shape of a bag. The stations at 7 km and 11 km have the lowest levels of dissolved oxygen. These stations are under heavy anthropogenic pressure from solid waste, air pollution, and the activity of microorganisms (depletion of the oxygen medium due to the degradation of the abundant organic matter in these stations), but are especially exposed by, for example, the heating of the water of these stations by discharges of raw sewage, rich in fermentable organic matter that can lead to microbial growth. Additionally, from 39 mg O2/l at the mouth to 236 mg O2/l at 7 km, the average COD values increased. This parameter's spatial evolution exhibits a bell-shaped curve (Fig. 4). In the center section of the estuary has the highest values, while the downstream and upstream sections have the lowest values. This discrepancy can be attributed to the middle section's several outflows and the absence of these outflows in the downstream and upstream sections. Similar to this, the substantial infiltration of seawater at the mouth of the Bouregreg estuary will undoubtedly dilution of the water and consequently a reduction in COD.

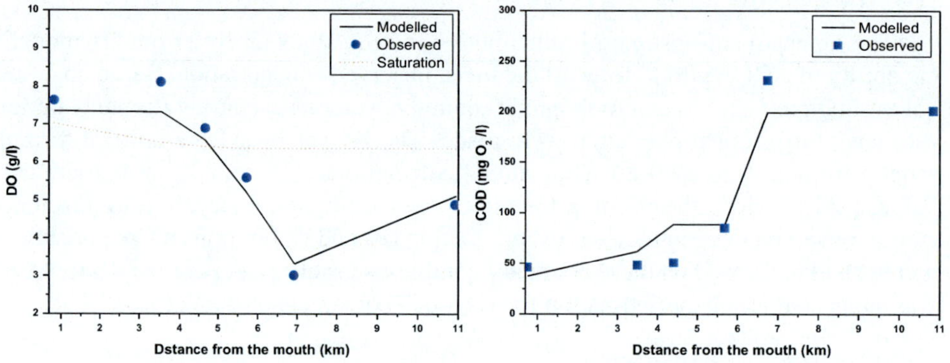

Fig. 4. Modelled vs observed of DO (mg/l not g/l) and COD along estuary axis.

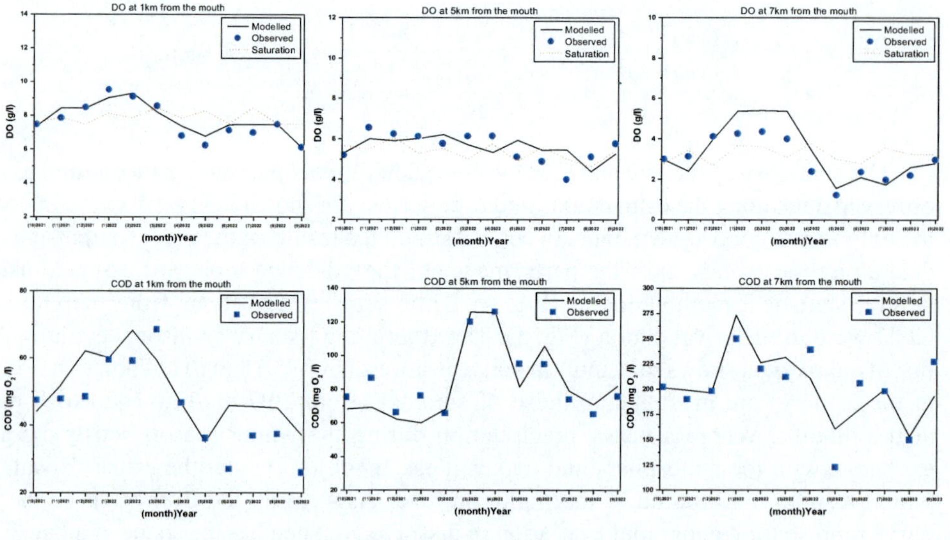

Fig. 5. DO(mg/l) and COD time series at Bouregreg estuary

In order to solidly achieve the model performance evaluation, various statistical analyzes were fully performed. These statistical results, comparing simulated and observed values, showed that AQUASIM is almost consistent in predicting water quality parameters of the Bouregreg estuary (Table 3).

Table 3. Statistical indicators of the hydrodynamic model performance

Statistical indicators		RMSE	NSC	R^2
Calibration (along estuary axis)	DO	0.08 (mg/l)	0.79	0.78
	COD	0.074 (mg O_2/l)	0.73	0.70
Validation (time series)	DO	0.077–0.084 (g/l)	0.720–0.820	0.740–0.800
	COD	0.060–0.680 (mg O_2/l)	0.690–0.730	0.700–0.750

5 Conclusions

The Bouregreg estuary has significant ecological importance in Morocco. Despite its ecological importance, the estuary is rarely studied and consequently ecosystem structure and function are rarely known. This study examined the evolution of DO and COD following extensive field observations and numerical modelling. The proposed model advances our understanding of ecosystem responses to human impacts. The results obtained in this paper will bring significant advances in understanding the ecosystem dynamics related to DO and COD in the Bouregreg estuary. There is a further need for research in the ecologically sensible modeling of the river water quality. This AQUASIM model could be employed in this endeavor during systems analysis at various degrees of complexity and human impact to enhance the practical comprehension of ecological processes in rivers and the identification of ecologically sound management approaches.

Declaration of Competing Interest. The authors declare having no conflicts of interest for this article.

Data Availability Statement. Data sharing not applicable to this article as no datasets were generated or analyzed during the current study.

References

1. Pritchard, D.W.: What is an estuary: physical view point. In: Lauff, G.H. (Ed.), Estuaries. No. 83, pp. 3–5. American Association for the Advancement of Science (1967)
2. Davidson, N., Council, N.C.: Nature Conservation and Estuaries in Great Britain. Nature Conservancy Council Peterborough (1991)
3. Savenije, H.H.G.: Salinity and Tides in Alluvial Estuaries. Elsevier, Amsterdam (2005)
4. Wei, X., Kumar, M., Schuttelaars, H.M.: Three-dimensional salt dynamics in well-mixed estuaries: influence of estuarine convergence, Coriolis and bathymetry. Journal of Physical Oceanography (2017)
5. Sarker, M.J., Tanmoy, M.H., Islam, M.S., Nazrul, K.M.S., Hossen, S., Ali, M.M.: Seasonal variation in the coastal water phytoplankton communities and their environmental responses at upstream and downstream of the steep Naf River in the south-western Bay of Bengal. Int. J. Aquatic Biol. **9**(5), 309–325 (2021)
6. Cherkaoui, E., Hily, C., Bayed, A.: Structure and distribution of intertidal benthic stands in an estuary of North Africa: the estuary of Bou Regreg. Marine Life **15**(1–2) (2005)

7. El Harim, A., Cherkaoui, E., Khamar, M., Nounah, A.: The impact of the depollution project on the Quality of the estuarine ecosystem of bouregreg (morocco atlantic coast). Int. J. Conserv. Sci. **10**(4), 763–772 (2019)
8. Haddout, S., Priya, K.L., Adarsh, S.: A predictive model for salt intrusion in estuaries applied to the Muthupet estuary (India) and Bouregreg estuary (Morocco). ISH J. Hydraulic Eng. **26**(4), 430–447 (2020)
9. Haddout, S., Priya, K.L., Samad, B.: Testing thermodynamics method for salt intrusion during infrequent astronomical event. Mar. Georesour. Geotechnol. **39**(10), 1139–1146 (2021)
10. El Amraoui, M., Salama, Y., El Amraoui, B., Lazrak, N., Mounkad, M.: Evaluation of the physicochemical and some trace elements of the surface waters of the moroccan atlantic estuary: case of the estuary of the river Bouregreg. Carpathian J. Earth Environ. Sci. **10**(2) (2015)
11. Priya, K.L.A., Haddout, S., Casila, J.C.C.: Implications of turbulent shear on clay floc break-up along the Atlantic estuary (Bouregreg) Morocco. Int. J. Sediment Res. **37**(2), 248–257 (2022)
12. Priya, K.L., Haddout, S., Adarsh, S.: Settling velocity of suspended sediments in Muthupet Estuary, India and Bouregreg Estuary, Morocco. In: Climate Impacts on Water Resources in India (pp. 125–146). Springer, Cham (2021)
13. Priya, K.L., Haddout, S., Adarsh, S.: Settling velocity of fine suspended sediments in Muthupet estuary, India and Bouregreg estuary, Morocco. In: Abstract of International Conference on Sustainable Technologies on Intelligent Water, pp. 16–19, February 2018
14. Kärnä, T., Baptista, A.M., Lopez, J.E., Turner, P.J., McNeil, C., Sanford, T.B.: Numerical modeling of circulation in highenergy estuaries: a Columbia River estuary benchmar. Ocean Model **88**, 54–71 (2015)
15. Elias, E.P., Gelfenbaum, G., Van der Westhuysen, A. J.: Validation of a coupled wave-flow model in a high-energy setting: The mouth of the Columbia River. J. Geophys. Res.-Ocean, 117 (2012). https://doi.org/10.1029/2012JC008105
16. Reichert, P.: AQUASIM 2.0: computer program for the identification andsimulation of aquatic systems. Dübendorf, Switzerland: Swiss Federal Institute for Environmental Science and Technology (EAWAG) (1998)
17. Association, A.P.H., Association, A.W.W.: Standard methods for the examination of water and wastewater. In: Stand. Methods Exam. Water Wastewater, p. 1000 (1995)
18. Khalifa, W.M.A.: Simulation of water quality for the El-Salam canal in Egypt. In: Presented at the water pollution 2014, The Algarve, Portugal, May 2014, pp. 27–37 (2014). https://doi.org/10.2495/WP140031
19. Khalifa, W.M.A.: Evaluation of water quality parameters using numerical modeling approach for the El-Salam Canal in Egypt. Int. J. Adv. Appl. Sci. **7**(2), 99–112 (2020). https://doi.org/10.21833/ijaas.2020.02.014
20. Legates, D.R., McCabe, G.J., Jr.: Evaluating the use of "goodness-of-fit" Measures in hydrologic and hydroclimaticmodel validation. Water Resour. Res. **35**(1), 233–241 (1999)
21. Zhang, P., et al.: Hydrodynamic and inundation modeling of China's largest freshwater lake aided by remote sensing data. Remote Sens. **7**(4), 4858–4879 (2015)
22. Owens, M., Edwards, R.W., Gibbs, J.W.: Some reaeration studies in streams. Air Water Pollution **8**, 469–486 (1964)
23. Wolf, P.: Simulation des Sauerstoffhaushaltes in Fliessgewässern. Vol. 53. Forschungs- und Entwicklungsinstitut für Industrie- undSiedlungswasserwirtschaft sowie Abfallwirtschaft (1974)
24. Thomann, R.V., Mueller, J.A.: Principles of Surface Water Quality Modeling and Control. Harper & Row (1987)
25. Greenberg, A.E., Trussell, R.R., Clesceri, L.S.: Standard Methods for the Examination of Water and Wastewater, 16th edn. APHA, Washington, DC, USA (1985)

26. Brahimi, A., Chafi, A.: Etude écotoxicologique de l'oued Za et de son affluent Oued Tizeghrane (Basse Moulouya, Maroc Oriental)[Ecotoxicological study of Oued Za and its tributary Oued Tizeghrane (Lower Moulouya, Eastern Morocco)]. J. Mat. Environ. Sci. **5**(5), 1671–1682 (2014)
27. Fischer, H.B., Liet, E., Koh, C., Imberger, J., Brooks, N.: Mixingin Inland and Coastal Waters. Academic Press, Cambridge, USA (1979)

Author Index

S. Haddout et al. (Eds.): CCORE 2022, SPEES, p. 191, 2023.
https://doi.org/10.1007/978-3-031-26967-7